ARCHITECTING RESILIENT SYSTEMS

WILEY SERIES IN SYSTEMS ENGINEERING AND MANAGEMENT

Andrew P. Sage, Editor

A complete list of books in this series appears after the index.

ARCHITECTING RESILIENT SYSTEMS

Accident Avoidance and Survival and Recovery from Disruptions

Scott Jackson

WILEY

A JOHN WILEY & SONS, INC., PUBLICATION

Published by John Wiley & Sons, Inc., Hoboken, New Jersey.
Published simultaneously in Canada.

Library of Congress Cataloging-in-Publication Data:

Jackson, Scott.
Architecting resilient systems : accident avoidance and survival and recovery from disruptions / Scott Jackson.
p. cm. — (Wiley series in systems engineering and management)
ISBN 978-0-470-40503-1 (pbk.)
1. Reliability (Engineering) 2. Systems engineering. 3. System safety. 4. Industrial safety.
I. Title.
TA169.J334 2010
620.001'171—dc22

2009018436

Printed in the United States of America

10 9 8 7 6 5 4 3 2 1

Finger in the Dyke

...the boy clambered up the heights until he reached the hole. His chubby little finger was thrust in, almost before he knew it. The flowing was stopped! Ah! he thought, with a chuckle of boyish delight, the angry waters must stay back now! Haarlem shall not be drowned while I am here!

—Mary Elizabeth Mapes Dodge,
Hans Brinker (1865, p. 158)

Contents

Preface

The story of the boy who stuck his finger in the dyke is, of course, a parable, that is, a story with a lesson. The incident did not happen, even in Holland, and besides, it is a physical impossibility. Anyhow, the lesson from the parable is that if you pay attention to small problems, then big problems are less likely to occur. All of this is relevant to the study of resilience.

So, how does this book differ from other writings on the subject? First, the science of resilience engineering is so new that there is little agreement about what it is. Most sources, for example, Westrum (2006a) list three aspects, as follows:

- The ability to prevent something bad from happening
- The ability to prevent something bad from becoming even worse, and
- The ability to recover from something bad before it becomes even worse than before.

Although some authors focus on the third aspect, this book takes the comprehensive approach and covers all three. Even though the first aspect can best be categorized as accident avoidance, it is difficult to separate it, its causes and its prevention, from the other two.

Second, there is not a consensus among the community about what a final definitive approach is to treating resilience. The approach discussed in this book is called the *systems approach*. Although this terminology is used in the literature, it is rarely described in detail. This book uses the comprehensive definition to include both holistic and analytical methods with the emphasis on the holistic, in which systems architecting plays a leading role.

The principles described in this book cannot be said to be completely new. They can be found, for example, in the fields of cognitive engineering, psychology, and socio-ecological science, among others. This book shows how these principles may be applied to a larger number of system types and environments than are typically shown.

Third, this book presents a framework for implementation that any organization, public or private, can use as a guide for establishing its own procedures and processes for achieving system resilience, which is the product of resilience engineering. This framework relies on an extensive familiarity with both the current research and also standard ways that programs are run in industry. Many conclusions will depart from conventional wisdom and accepted practices in industry and government.

Fourth, the principles presented in this book are intended to be applied across all domains in which accidents may occur. Case studies presented here include, for example, chemical facilities, nuclear power plants, civil infrastructure elements, commercial aircraft, and space vehicles.

Fifth, this book presents resilience within a framework that consists of the following three primary enabling elements: capabilities, culture, and infrastructure. It is not the intent of this book to say that this is the only way to structure system resilience or even the best way to structure it; it is just a convenient and logical way. The characteristics of each of these elements are found in the existing literature. This book only attempts to put these characteristics into a logical and understandable order to help implement them.

Another logical question is whether this is a management or a technical book. The answer is "yes" to both. A point that all resilience engineers agree on is that system resilience depends on both management and technical processes and the interaction between them. It is not possible to have a grasp of system resilience without understanding both management and technical aspects.

Finally, the question should be asked: To whom is this book directed? In truth, it is difficult to identify anyone who would not gain something useful from it. First, students are important because they may be the ones facing the difficult decisions in the future. Because this book is intended as both a textbook and a book of general interest, appreciation by the student audience is critical. Another particular audience is anyone in so-called high-consequence endeavors in which lives are at stake. These endeavors include nuclear power plants, commercial aircraft, and public agencies whose role is to plan for natural and human-made disruptions, such as terrorist attacks, hurricanes, and earthquakes. In addition, hospital administrators will find much of interest. A special audience is the systems architect. These people are responsible for determining the arrangement of their system, be it an organization, a software code, or a hardware system. Everyone in these endeavors will find value in solving the cultural paradigm challenge discussed in Chapter 5, Culture. The systems architect and the designer will find value in the resilience attributes outlined in Chapter 8, Resilience Architecting. These are only a few examples; it is hoped that all who read this book will find something of value.

For all of the above reasons, I hope this is a useful and timely book.

SCOTT JACKSON

Los Angeles
April 2009

Acknowledgments

The following advisors have provided invaluable support for this book:

From the University of Southern California are Azad Madni, PhD (resilience and cognitive science); and James Moore, PhD (railroad systems).

Others include Katherine Erlick, PhD (organizational psychology); Stuart D. Hann (safety); Daniel Kelly, James Lambrechts, and Robert Costa, San Francisco Fire Department (fire protection infrastructure); Shaun A. Jackson, PhD, Weber State University (bibliographic assistance); and my wife Carole (general editing).

Notes on Terminology

The following notes are intended to discuss some of the terminology used in this book. Because of the interdisciplinary nature of resilience, the terminology used in various disciplines in this book will vary. Some of these disciplines are resilience engineering, system safety, cognitive engineering, systems engineering, and others. The terminology used will most often reflect the preferences of the systems community, that is, those whose endeavor is to define human or technological systems using the systems approach, including both analytic and holistic methods described in this book.

There is no intention to preempt any of the terminology found in other domains; it is solely to place the content in a context familiar to the systems community.

ARCHITECTING

Architecting is the arrangement of the parts of a system, that is, the architecture of a system. This is a word not often found in the resilience literature. Maier and Rechtin (2009, p. 423) define an architecture as "the structure – in terms of components, connections and constraints - of a product, process, or element." Maier and Rechtin (2009, p. 423) also define architecting to be "the process of creating and building architectures . . ." However, Maier and Rechtin reveal the differing views of architecting by saying that it "may or may not be considered a separable part of engineering." In this book, the broader interpretation is used, that is to say, that architecting considers all aspects of defining the structure of a system, both analytic and holistic.

CAPACITY, ABSORPTION, AND MARGIN

The word *capacity* does not frequently occur in the resilience literature. However, it is implied by the two concepts. *absorption* and *margin,* discussed by Woods (2006b). For convenience, these two concepts have been combined into a single category called capacity. Apart from the convenience, it was felt that this is a logical categorization.

HEURISTIC

The term *heuristic* means a design principle, guideline, or rule that has been learned from experience, especially with respect to the definition of the architecture of a system. Maier and Rechtin (2009, p. 424) define heuristic as "a guideline for architecting, engineering or design." This term is widely used in the systems community, that is, among experts in systems architecting, such as Rechtin (1991). This term is not often used in other communities, such as the resilience engineering, systems safety, and cognitive engineering communities. Billings (1997), for example, provides a list of "requirements" for the relation between humans and automated systems. In the context of this book, these requirements are more appropriately called heuristics.

HOLISTIC VS. ANALYTIC METHODOLOGIES

There is much in the literature regarding the difference between holistic and analytic methodologies. Checkland (1999, p. 15) describes the basis of "systems analysis" as the assumption "that problems can be solved by making a choice between alternative means of achieving a known end." Checkland ascribes systems analysis as being more appropriate to and more successful at defining systems, which he calls "hard systems," for which the system's need is "given" (p. 191). These approaches have been applied extensively to technological systems, such as aircraft and space vehicles. If the end result is less predictable, Checkland calls these systems "soft systems." Systems containing human components fall into the latter category. Checkland says that the results for applying analytic methodologies to soft systems have been "disappointing" (p. 15).

According to Checkland, analytic thinking has its roots in scientific thought as articulated by Descartes. According to Checkland (p. 46), Descartes believed that all complex problems could be solved by dividing "each of the difficulties . . .into as many parts as might be possible and necessary in order best to solve it." Checkland says that this process is called "analytical reduction" and is the basis for analytic thinking.

The opposite of analytic thinking is holistic thinking. Holistic thinking considers a system as a whole rather than as simply the sum of its parts. One feature of holistic thinking is that it allows for the system to demonstrate "emergent" properties, that is, properties that cannot be predicted by examining the individual parts. In this book, resilience itself is viewed as an emergent property. Checkland (p. 13) says that the "systems paradigm is concerned with wholes and their properties." However, Checkland (p. 74) predicts "systems thinking and analytical thinking will come to be thought of as the twin components of scientific thinking." It is this prediction that forms the basis for the assertion in this book that both holistic and analytic methodologies are necessary to create resilient systems.

However, it is not the intent of this book to make a hard distinction between holistic and analytic methodologies. Some methodologies may be either holistic or analytic depending on the context. One simple test for whether a methodology is holistic or analytic is whether it addresses multiple components and their relationships.

For example, interface analysis, discussed in Chapter 6, Capabilities, is described as analytic. Since all interface analyses, by definition, consider at least two components, these analyses might be considered holistic. The distinction is as follows: If the analysis only determines how the components "fit" together and have compatible requirements, then the process can be considered analytic. On the other hand, if the analysis has a broader goal of determining how the components function together, then the process can be considered holistic. So, the lesson is as follows: Do not consider the categories assigned in this book to be absolute or applicable to all situations.

The holistic methodologies discussed in this book include architecting through the use of heuristics, and other considerations, such as risk and culture. However, holistic methodologies are not limited to these topics. They might include, for example, simulations that examine various system constructs and their goals. Through these simulations, emergent behavior may be evident.

INTERELEMENT COLLABORATION

The term *interelement collaboration*, which is discussed in Chapter 8, Resilience Architecting, was introduced as a concept to expand on the term *cross-scale connectivity* discussed in the literature, for example, in Woods (2006b). Interelement collaboration does not replace cross-scale connectivity but rather was intended to add an extra dimension to it. Although cross-scale connectivity emphasizes the connections between elements of a system, inter-element collaboration adds the dimensions of coordination and cooperation among the elements.

RESILIENCE AND RESILIENCE ENGINEERING

Both *resilience* and *resiliency* can be found in the literature. However, most authors in the resilience community have agreed that these words are essentially the same and have chosen to use *resilience*, as has been done in this book. As for *resilience engineering*, this is a valid phrase assuming that the word *engineering* is used in a broad sense as Checkland (1999, p. 10) notes regarding systems engineering, below. However, in this book, the term *resilience engineering* is used only when referring to the book by Hollnagel et al. (2006) and to the Resilience Engineering Network.

SYSTEMS ENGINEERING

The term *systems engineering* is used sparingly in this book because of the various interpretations of this phrase. The traditional view of systems engineering is that it was a reductionist, analytic methodology that focused on individual elements of a system. Traditionally, systems engineering was thought applicable only to technological systems. However, despite its reductionist reputation, systems engineering has always, to some extent, incorporated holistic aspects as described above.

Today this term has come to embrace all methods used to define a system, both analytic and holistic. Systems being addressed include all the systems discussed in Chapter 2, System Resilience and Related Concepts, including technological and human systems. In short, it is equivalent to applied system science. Checkland (1999, p. 10) says that the *engineering* in systems engineering should be interpreted in a broad way, for example, to "engineer a meeting or a political agreement." However, for the purposes of this book, the terms *systems approach*, *analytic approach*, and *holistic approach* are used because there is more of a common understanding of their meaning.

Chapter **1**

On Resilience

The world suffered vicariously as firefighters entered the towers and never returned. We imagined ourselves as passengers on United 93 as it plunged toward a field in Pennsylvania. We felt the temperatures rise as the Space Shuttle Columbia's tiles peeled away. In New Orleans, we saw waters break through the levees and flood our houses while we waited in vain for help to come.

And yet, the persistent question remains: could all these calamities have been prevented; or worse, would it have been possible to survive and recover from them and continue functioning? And furthermore, whose fault was it? Was it bad design, or a cultural problem, or management, or politics, or something else? The answer is "yes" to all these questions. And the approach to disaster avoidance, survival, and recovery from such disruptions requires that expertise from a multitude of disciplines be executed to an unprecedented degree. Hence, these three elements, accident avoidance, survival, and recovery constitute what has come to be called resilience. These three elements will be discussed more in Chapter 2, System Resilience and Related Concepts.

The opposite of resilience is called brittleness. One challenge of research is to determine ways to find out when a system is "drifting" toward brittleness. What can be measured? What can be observed? And, most importantly, what can be done to stop the drift toward brittleness?

The purpose of this book is to answer these questions within an integrated framework. To create this integrated framework, many factors are brought together: management and technical functions should be considered a single interconnected discipline. Other disciplines, such as organizational psychology, are brought into the mix. The distinctions among human, software, and

Architecting Resilient Systems: Accident Avoidance and Survival and Recovery from Disruptions.
By Scott Jackson

hardware systems are erased. Chapter 7, Infrastructure, defines an infrastructure system that encompasses all relevant organizations. Finally, to create systems that are resilient to and survive major disruptions requires considerations far beyond current practices that might be considered just good engineering. This framework extends beyond the breadth currently envisioned in most academic, government, and industrial institutions. However, the expectation is that these practices will be incorporated into all aspects of system development wherever they may be needed.

1.1 THE MULTIDISCIPLINARY CHALLENGE

We live in a world of specialists. Even the family doctor has a limited understanding of what the cardiologist does. There are very few Renaissance men like Leonardo. But even more to the point, the gulf among artists, psychologists, managers, and scientists is enormous. Resilience demands Renaissance men and women who are at once practitioners of multidisciplinarity and transdisciplinarity. The multidisciplinary nature of system resilience makes it clear that system resilience is not another specialty but rather a collaborative effort among many specialties, both technical and nontechnical.

Nicolescu (2007, Part 1: The war of disciplines) describes multidisciplinarity as using several disciplines to enhance the topic in question. The bringing together of management and technical capabilities satisfies this definition. Transdisciplinarity, however, according to Nicolescu, "concerns that which is at once between the disciplines, across the different disciplines, and beyond all disciplines." Performing the risk management process and addressing the psychological paradigm of risk denial seem to fall in this category. Jackson (2007) focuses on the multidisciplinary aspects of system resilience.

Researchers all over the world, such as the Resilience Engineering Network, are studying system resilience. Hollnagel et al. (2006) and Hollnagel and Rigaud (2006) are two examples of the products of the Resilience Engineering Network. They are a unique group, which consists of engineers, psychologists, sociologists, physicians, and other specialists. They all have one thing in common: an interest in solving the mysteries of resilience and a new discipline of resilience engineering. Could Chernobyl have been prevented? If not, how could anyone survive such an event? These experts all agree on one point: that the answer to this question does not lie solely in engineering expertise but rather in a broad range of interconnected disciplines, both technical and human. In addition, the study of resilience has attracted the attention of the legal profession. The George Mason School of Law has instituted the Critical Infrastructure Protection (CIP) Program. The CIP report (2007) provides an in-depth study of the benefits of a resilience approach *versus* the traditional protection approach.

These researchers come with many ideas, and there is not yet a consensus among them on many points. This book will summarize many of those ideas.

The one principle that the researchers all agree on is that resilience to disaster and survival of disruptions are not purely technical subjects. In the early days of space flight, many failures could be traced to technical causes. That is, a system would fail because of unreliable components. Today, such failures are largely under control. However, systems continue to fail because of causes beyond the technical. The Caltech physicist Richard Feynman (1988) had asked National Aeronautics and Space Administration (NASA) management what they thought the probability of failure was for a space shuttle. They replied, relying on reliability analysis, that it was about 1 in 100,000. Working engineers put the number at about 1 in 100. Range safety experts estimated the number to be 1 in 25. History has shown that the latter numbers were closer to the truth. In short, the disparity in these estimates demonstrates that Feynman seemed to understand that the causes of catastrophe were far beyond technical considerations, at least as currently understood in the engineering community.

1.2 THE CONCEPT OF THE SYSTEM

Whenever two or more things act together to achieve a common purpose, you have a system. A mousetrap is a system; the government is a system; a troop of Girl Scouts is a system; and a chemical power plant is a system. If all the parts of the system do not work properly, the whole system may fail. The concept of the system is discussed in Chapter 2, System Resilience and Related Concepts.

1.2.1 The Paradox of Humans in the System

To understand and design for resilience, the role of the human needs to be understood. Humans are not simply operators of the system, like pilots. Nor are they simply maintainers of the system, like aircraft mechanics. They are not only producers of the system, like factory workers. They are not simply designers of the system, like engineers. All the above examples are parts of the system. Sometimes humans constitute the entire system itself, such as a troop of soldiers. These are called human systems. Some human systems, such as hospitals, have hardware and software components, such as X-ray machines and other test equipment. However, the predominant elements are human, such as doctors, nurses, and other staff members. For these types of systems, the term "human-intensive systems" applies.

Many researchers, such as Bennis (1999), have studied humans within an organizational context. However, the design rules of human systems can best be determined by the attributes, laws, and heuristics discussed in Chapter 8, Resilience Architecting. Heuristics are the design rules of systems architecting, as described by Rechtin (1991), rather than verifiable requirements as in the traditional systems approach. There is no way to test the humans in all possible scenarios to determine whether they perform correctly. There are ways to reduce human error, such as training. Nevertheless, human actions may be highly unpredictable leading to unpredictable outcomes. These outcomes are a result of Type B disruptions that is, disruptions caused by degradation of function, capability, or capacity, which will be discussed later in Chapter 3, Disruptions. The saving grace of humans is that they are adaptable and can sometimes create solutions not even imagined by the designers. An example is the restoration of electrical power in New York after the attack on the twin towers as described in Chapter 4, Case Histories.

In any discussion of humans, the question of *human error* always arises. This is a subject about which there is abundant misunderstanding. This issue is particularly important in the health care domain as described by Rooney et al. (2002). The authors point out that "the majority, 80 to 85%, of human errors result from the design of the work situation, such as the tasks, equipment and the environment." These data exclude malevolent acts. In other words, human errors are *systemic* in nature and cannot wholly be blamed on individuals. Reason (1990, p. 173) makes the following statement with regard to human error:

> Rather than being the main instigators of an accident, operators tend to be the inheritors of system defects created by poor design, incorrect installation, faulty maintenance, and bad management decisions. Their part is usually that of adding the final garnish to a lethal brew whose ingredients have already been long in the cooking.

Chapter 3, Disruptions, describes human error as a source of disruptions, that is, events that can lead to disaster. Hence, the inevitable conclusion is that the pilot of the aircraft in the Nagoya incident described in Chapter 4, Case Histories, might not have made the fatal mistake if the appropriate features had been designed into his aircraft. Chapter 8, Resilience Architecting, describes these features as adaptable and agile. Nevertheless, Rooney et al. (2002) conclude that significant reductions in human error can be achieved with good training, good work situation design, good procedures, and a good corrective action system. Rooney et al. (2002, p. 35), discussing human error in the health care domain, also recognize the adaptability of the human by insisting that a prerequisite for reducing human error is the "freedom to act."

In short, whereas humans may be a major source of accidents, designers of most modern systems, such as commercial aircraft, recognize that the adaptability of humans makes them essential components for resilience.

1.2.2 The Infrastructure as a System

An infrastructure is the collection of people and equipment that design, operate, produce, and maintain a product system, such as an aircraft or spacecraft. An infrastructure also consists of the relationships between these elements. Or the infrastructure itself, such as a civil infrastructure, may be the system. Thinking of this collection as a system is not a traditional way to characterize a system. Yet it is central to resilience. One has to think of what the pilot does, what the maintainer does, and so on. It is this system that should operate flawlessly so that the aircraft or spacecraft can also operate flawlessly. This concept becomes tractable when the boundary of the system is defined. Chapter 2, System Resilience and Related Concepts, describes the systems approach in which boundary definition is essential. Because all these human elements contribute to the objective of the system, it is only logical to conclude that they are components of the system.

The scope of the infrastructure system becomes even more foreboding when the number of people and organizations involved are considered. The communications among these organizations become points of brittleness. The people themselves and the decisions they make are critical to the system success. For example, a patient waits for a donated heart. Is the blood type correct? An error in this situation is a matter of life or death. This is an example of a fragile infrastructure system consisting of a hospital, doctors, a blood donation organization, and other elements. The concept of the infrastructure system is discussed more fully in Chapter 2, System Resilience and Related Concepts, and in Chapter 7, Infrastructure.

1.2.3 The Architecture of a System

So what is it that makes a system resilient? For the most part, it is its architecture. The architecture of a system is how the parts of the system are arranged and how they interact with each other. Rechtin (1991) coined the term "architecting" to describe the process of creating an architecture. For example, if a fire protection system has multiple ways of putting out fires, then those methods would be part of the fire protection system's architecture. This example appears in Appendix A, Domain-Specific Example for Architecting a Fire Protection Infrastructure System.

Zachman (2007) writes on the subject of architecture, especially enterprise architecture. Zachman (2007, p. 6) says:

> "Architecture" is the set of descriptive representations relevant for describing a complex object (actually any object) such that an instance of the object can be created and such that the descriptive representations serve as the baseline for changing an object's instance (assuming that the descriptive representations are maintained consistent with an instantiation).

Zachman notes that many people confuse the *implementation* of an architecture, for example a building, with the architecture itself, which Zachman describes as a representation. Zachman's definition is valid for any type of system, for example, infrastructure systems or technological systems.

In Zachman's definition, descriptive representations are models that enable the analyst to view the object or system from a desired perspective. It can be said that an architecture is an abstract view of a system and not a physical view as observed in a photograph. One of the most common architectural views is the hierarchical representation, but there are many other views.

So how does one "architect?" Rechtin suggests a set of heuristics, or guidelines derived from experience for creating the architecture of a system. Chapter 8, Resilience Architecting, provides heuristics focused on creating a resilient system.

1.3 DISRUPTIONS

When Hurricane Katrina slammed into New Orleans, the effect was a *disruption* of the normal activities of the city. When the power went out on Apollo 13, the loss of power was a disruption of the mission of that space vehicle.

The analysis of disruptions is essential to the study of resilience. The question is as follows: When the function of a system is disrupted, will it cause a catastrophic failure? Can the systems survive the disruption? Can the system recover and continue to function at any level? To what extent is the resilience of the system dependent on the type of disruption?

When the Tacoma Narrows bridge collapsed in 1940, the failure can be categorized as a Type A, or *disruption of input*, as defined in Chapter 3, Disruptions. That is to say, the bridge experienced a phenomenon not known to the designers, namely the effects of the turbulent aerodynamic boundary layer on the bridge.

When the Challenger spacecraft failed, it experienced a degradation of function, capability, or capacity, which is the second type of disruption. These disruptions are called systemic disruptions. The O-rings, which are an internal component of the system, failed, resulting in the catastrophe. When a human is part of the system and makes an error, as in the Nagoya aircraft failure, this error is an error of function.

Hurricane Katrina is an example of a disruption caused by a change in the environment, which is also a disruption of input. However, the failure of the levees can also be categorized as an internal, or systemic, disruption. The winds of Hurricane Katrina constituted a disruption of immense magnitude. Hurricane Katrina is also an example of a network disruption in which the failure of one element of the system, for example, the levees, resulted in the failure of other elements, for example, the transportation system. All these examples are discussed in Chapter 4, Case Histories. These examples also show that the presence of humans and software in systems introduce unprecedented degrees of disruption.

Sometimes disruptions of function can result from the interaction of two elements of the system. This kind of disruption can occur even when the individual components operate as designed. Such was the case for the Mars Polar Lander, in which the interaction between the landing struts and the software resulted in the premature shutdown of the engines. Although such disruptions can be attributed to poor integration, predicting them may require analysis to a greater level of detail than is common.

1.4 ADAPTABILITY

Adaptability can be said to be an *emergent* characteristic of a system that enables it to avoid, survive, or recover from a disruption. Many existing and past systems had a large degree of natural, or built-in, adaptability, discussed in Chapter 8, Resilience Architecting. Apollo 13 was a good example as discussed in Chapter 4, Case Histories. When the main power failed, the crew saved power by moving to a smaller module. In this way, they were restructuring the system, which is a key attribute of adaptability. Adaptability is, in this case, emergent because it illustrates the relationship between the modules and is not a characteristic of each module when treated singly.

Another attribute is interelement collaboration or communication and cooperation between elements. Hurricane Katrina is an example in which virtually no communication or cooperation occurred among government agencies. This system failed the adaptability test with catastrophic results. The rapid restoration of power in New York after the attack on the twin towers is an example of adaptability. This case is discussed in Chapter 4, Case Histories. Adaptability can be designed into a system using the holistic methods that are part of the systems approach as described in Chapter 8, Resilience Architecting. As explained in Chapter 2, System Resilience and Related Concepts, holistic methods take into account the relationship among system elements, whereas analytic methods focus on individual components.

1.5 CULTURE

The Columbia Accident Investigation Board (2003) found that cultural factors were at least as important a contributor to the Columbia disaster as technical factors. The NASA culture of accepting risk in the pre-Challenger days was described by Vaughn (1996, p. 415) as the "normalization of deviance." She uses this phrase to describe a cultural environment in which risks were accepted to be an established norm.

The term *culture*, as discussed in Chapter 5, Culture, may refer to the individual, organizational, or national beliefs that govern our actions. Although the cultural influences on resilience are well documented, for example, in Vaughn (1996) and in the Columbia Accident Investigation Board report (2003), this aspect has received little attention in industry or government.

Basically, two ways are available for an enterprise to address culture. First, designers of the enterprise can make their processes so rigorous that these processes would be virtually impervious to culture. Alternatively, they could attempt to change the culture. Neither of these techniques is easy. Chapter 5 outlines some potential methods for changing culture.

1.6 MEASURING RESILIENCE

A major topic among resilience researchers is whether resilience or, better put, lack of resilience, called *brittleness*, can be predicted or measured. One school of thought is that accidents are so random that any type of prediction is out of the question. However, others point to defects and near-misses that almost always precede major accidents. This idea is called the iceberg theory, as discussed in Chapter 10, Measuring Resilience. Can it be said, for example, that an aircraft that requires more maintenance is more likely to have an accident? The answer to this question is not known, but it is possible. Some statistical evidence exists for example, from the Scottish Railways study, which collected and analyzed data on defects and major accidents in the railway domain. This study indicated a statistical correlation between minor events and major accidents, as described by Wright and Van der Schaaf (2004).

If developed, such evidence would have an enormous benefit. This information could be used to create better designs, better operational procedures, or better maintenance procedures. In short, although the iceberg theory does not directly measure resilience, it could be a step in the process of designing more resilient systems.

Otherwise, the following conclusions can be drawn. First, traditional reliability and safety analyses can be used to arrive at quantitative results. However, these results are based almost entirely on historical data and may reflect, to a certain extent, the ability of the system to survive predicted disruptions. Second, data exist for human error in some domains—for example, in commercial aircraft. Hence, to the extent that such data do exist, they can be used to supplement reliability and safety analyses to arrive at quantitative results. Finally, there is the issue of unpredicted threats, as discussed in Chapter 3, Disruptions. It can be concluded that the measure of the resilience to unpredicted threats is only possible to the extent that a given resilience method has proven useful against *other* unpredicted threats.

1.7 THE CHALLENGES

If there is a more difficult problem to solve than resilience, it is hard to say what it is. Can multiple disciplines as widely diverse as engineering and psychology, for example, be corralled to analyze resilience in its entirety? Can multiple organizations, including, for example, aircraft developers, customers, government agencies, operators, and maintainers, be integrated to solve communications and decision-making problems? Can disruptive events, such as conflicts between operators and software, be predicted and eliminated? Can adaptive systems be created to survive and recover from major disruptions, such as terrorist attacks and hurricanes? Can cultural change actually be achieved that will result in risk-conscious organizations that will anticipate and address threats to resilience? Finally, are there indicators of potential weaknesses in systems that can be exploited to make more resilient systems? All of these examples are potential steps toward designing resilient systems.

1.8 FURTHER EXPLORATION

Given what you know so far about system resilience, write a short essay on all its aspects. These aspects should include resilience itself, the importance of multiple disciplines, the concept of a system and its broader meanings, the concept of an infrastructure system and its importance, adaptability, and resilience measurement. Take advantage of outside sources for this essay.

Chapter **2**

System Resilience and Related Concepts

System resilience is all about the processes, disciplines, and infrastructure that need to be in place to make sure that such events as Challenger, Columbia, Chernobyl, Piper Alpha, and Bhopal never happen again, how survival from such disruptions may be achieved, and how each system may continue to operate. Many excellent books have been written on the causes of these accidents. Vaughn (1996), for example, has written an exhaustive analysis of the Challenger accident. Paté-Cornell (1990) published the ground-breaking paper on the North Sea oil platforms. Reason (1997), Leveson (1995), and Chiles (2002) also analyzed many such accidents. Hollnagel et al. (2006) go a step further; they present a set of concepts and precepts that form the attributes of a resilient system, that is to say, a system that is not prone to accidents and will survive major disruptions and continue to function. The focus of this book, however, is *implementation*. That is to say, how does a government agency, a developer, an operator, or a maintenance organization put into place an organizational and technical system to address the root causes identified by these authors?

One conclusion that these authors have in common is that major accidents cannot be attributed solely to technical processes. In fact, few major accidents can be attributed solely to reliability failures, that is, the failure of some critical component to function in accordance with the predicted mean time between failures (MTBFS). The causes encompass an array of organizational, managerial, human, and psychological topics.

Architecting Resilient Systems: Accident Avoidance and Survival and Recovery from Disruptions.
By Scott Jackson

Hence, one theme of this book is that the system resilience planner, whether in a governmental or private organization, should create a process infrastructure that goes far beyond the traditional technical disciplines. Furthermore, this process infrastructure should be integrated with the technical processes into a single entity and not partitioned into separate pieces. Another theme is that the organizations involved—governmental, developmental, operational, and support—should also be integrated into a single entity. Although the obstacles, contractual and other, to this goal may seem insurmountable, they are both necessary and achievable.

An understanding of some key concepts is necessary to begin the implementation of system resilience. A broader understanding of such terms as *system* and *infrastructure* will aid in this understanding, as discussed later in this chapter.

2.1 RESILIENCE

Resilience, as described by Westrum (2006a), consists of at least two of the following:

Avoidance—This term is defined to describe the preventive aspects of system resilience in response to a disruption, either internal or external as defined in Chapter 3, Disruptions. Although many writers do not include avoidance within the scope of resilience, other writers do. Because the capabilities required to avoid an accident, for example, flexibility, are the same as those required for recovery, this book includes both aspects. In the resilience context, avoidance goes beyond traditional system safety considerations to consider the *anticipation* of an accident based on the ability to detect "drift" toward brittleness and possible accidents. Common methods to detect drift are sensors, cameras, and human observation.

Survival—This term simply implies that the system has not been destroyed or totally incapacitated and continues to function when experiencing a disturbance.

Recovery—This term is used to pertain to the capability of surviving a major disturbance with reduced performance. This capability is a focus of system resilience. Again, it is only one of the three aspects of resilience, namely accident avoidance, survival, and recovery.

One might say that if avoidance is achieved, then survival and recovery are moot; that is, they are automatically achieved. However, because some systems are so complex and unpredictable, avoidance may not be achievable, and we must rely on survival and recovery to save the day.

2.1.1 The Three Phases of Resilience

The three aspects of resilience above can be viewed as occurring before, during, and after a disruption. Richards et al. (2007) refer to these phases as epochs. Table 2.1 describes these phases. During each phase, steps can be taken to deal with the disruption. These steps can only be taken if the system in question has

Table 2.1. The Three Phases of Resilience

Phases	Name	Epoch[a]	Description
1	Avoidance	1a	Predisruption
2	Survival	2	Loss of capability during disruption
3	Recovery	1b	Recovery of capability after disruption

[a] Richards et al. (2007).

certain attributes described in Chapter 8, Resilience Architecting. Chapter 8 also describes heuristics, or design rules, that will be most effective in each phase.

During Phase 1, the system will take steps to avoid any damage at all from the possible disruption. During Phase 2, the system will attempt to minimize the damage caused by the disruption. During Phase 3, the system will take steps to recover as much capability as possible.

So, the creation of resilience is the whole purpose of the enterprise focused on achieving system resilience. System resilience cannot be achieved without a resilient enterprise system to enable it. For systems such as civil infrastructure systems, the resilient enterprise is the infrastructure system itself that is being disrupted by, for example, a hurricane. For product systems, such as spacecraft, the resilient enterprise is the management infrastructure that enables the resilience of the product. Resilience is also the ability of organizational, hardware, and software systems to mitigate the severity and likelihood of failures or losses, to adapt to changing conditions, and to respond appropriately after the fact.

Many sources, for example, Woods (2007), refer to the opposite of resilience as brittleness. According to Woods, a brittle system cannot adapt to the forces of an unanticipated disturbance and breaks down under the stress.

Most authorities focus on the preventive aspects of resilience. Hollnagel (2006, p. 14) states that "resilience is the ability to maintain effective barriers that can withstand the impact of harmful agents and the erosion that is a result of latent conditions." To this end, this book focuses on the preventive aspects, namely, the capabilities that have to be in place, the culture that has to exist, and the infrastructure within which a system is developed, deployed, operated, and maintained. If these factors are addressed in advance, then these steps can be considered to be preventive. Thus, even recovery aspects can be considered preventive because they are addressed in advance.

But, once again, it is not just the product system that is resilient. It is the entire infrastructure system. Weick and Sutcliffe (2001, p. xiii) refer to this type of organization as a "high reliability organization (HRO)." Product system resilience may be measured by reliability, but the entire enterprise resilience is much harder to define and to measure. That is what this book is all about, namely, that the processes for creating resilience in systems go beyond traditional methods. Human-intensive systems, however, have no product. Regional infrastructures and governments are examples. These types of systems require resilience. They, too, are the subject of this book.

2.1.2 Protection versus Resilience

Some authors distinguish between resilience and protection. These authors describe protection as the traditional way of planning for disruptions; protection is generally fixed and inflexible. Resilience is more adaptable and can recover from the disruption. In the terms of this book, protective systems are more brittle. According to McCarthy (2007, p. 7),

> "Protection" includes "protective measures," which refer to actions, procedures, or physical impediments used to mitigate vulnerabilities, minimize consequence, and reduce risk. Simply put, protective measures are implemented to defend against harm to property, personnel, or mission execution. Examples of protective measures include but are not limited to, the following: surveillance cameras, security patrols and response capabilities, fencing, employee and visitor credentialing, and intrusion detection systems.

McCarthy's examples are from the civil infrastructure domain.

Protective systems generally cannot handle many kinds of disruption or respond in multiple ways. Scalingi (2007, p. 58) points out that protection is not separate from resilience. She says, "Resilience *includes* protection and prevention, which are important elements of it." In other words, it is possible to architect a system that protects against a disruption and also is adaptable enough to recover from it.

Another way to understand the difference between protection and resilience is to describe these systems as either tightly coupled or loosely coupled. Protective systems are tightly coupled, and resilient systems are loosely coupled.

According to Pommerening (2007, p. 11), "Tight coupling is associated with time-dependent and invariant processes with deliberate buffers, and little, if any, slack. Resources cannot be substituted easily or at all. Loose coupling means output delays are possible and the order of processing can be changed. In such systems, alternative methods and redundant resources are available, and fortuitous buffers and substitutions are possible." From these descriptions, it is possible to see the resilience attributes of flexibility, tolerance, and interelement collaboration in loose coupling, as well as the attribute of capacity in tight coupling. These attributes are discussed in Chapter 8, Resilience Architecting.

Grote (2006) also discusses loose coupling in organizations and how organizational rules can enable loose coupling. Grote discusses goal rules, process rules, and action rules. Her thesis is that maximum loose coupling, and therefore flexibility, can be achieved through goal rules.

2.1.3 System Resilience Concept Diagram

Figure 2.1 poses a basic concept diagram for system resilience. A concept diagram, in this instance, is a description of the entities that make up resilience and the relationships among these entities. The key features are its three

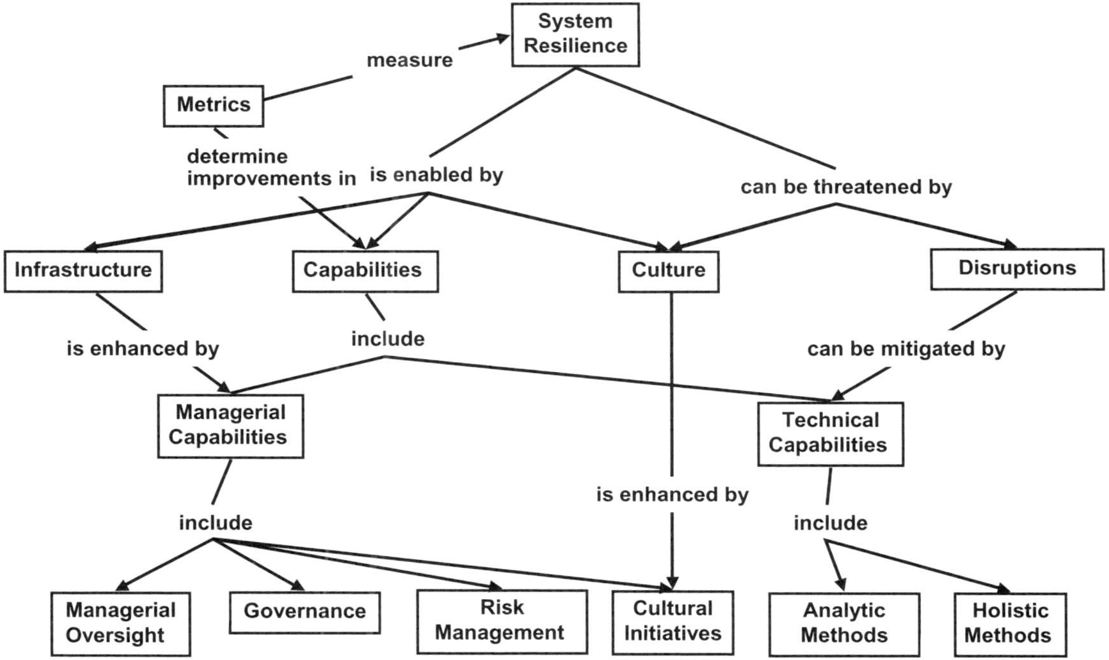

Figure 2.1. A concept diagram for system resilience.

principal enabling entities: culture (Chapter 5), capabilities (Chapter 6), and infrastructure (Chapter 7). This figure also shows that two phenomena threaten resilience. First, resilience can be threatened by culture, the main characteristics of which are the negative paradigms that exist within a culture and jeopardize resilience. Second, resilience can be threatened by disruptions; these events jeopardize the basic functionality of the system, as discussed in Chapter 3, Disruptions.

This figure provides a roadmap to this book because each element is treated within it, either as a separate chapter, a section of a chapter, or a diagram. Chapter 6, Capabilities, describes the technical and managerial capabilities that are required. This feature underlined the theme that resilience does not just consider technical factors but also managerial factors.

2.2 SYSTEM

Before we can discuss the concept of infrastructure on Figure 2.1, the concept of the system needs to be understood. According to the International Council on Systems Engineering (INCOSE) (2006, p. Appendix-8), a system is "a combination of interacting elements organized to achieve one or more stated purposes." For our purposes, systems can be divided into these broad categories: technological systems, technological systems with human components, product-centered infrastructure systems, human-intensive systems, process systems, systems of systems, and socioecological systems. In this book, we will address only human-made systems or natural systems in which there is human intervention. Natural systems, such as the solar system and flocks of geese, will not be covered. Systems of ideas, which are also called conceptual systems by Blanchard and Fabrycky (2006, p. 7), are potential for analysis because they are human made. Organizational systems are conceptual systems, and therefore are human made. Table 2.2 shows types of systems and example application areas that will be discussed in the following paragraphs.

2.2.1 Technological Systems

Technological systems are those systems whose focus is on hardware and software. Of course, spacecraft and aircraft fall into this category. However, it also includes nuclear power plants, off-shore platforms, and others. Most technological systems also have human components. Those aspects are covered in the next section of this chapter.

However, technological systems have been the domain of the analytic approach for many years. This approach cannot be abandoned, but it is only the beginning; it is not sufficient for system resilience.

Chapter 6, Capabilities, lists many of these aspects, for example, requirements analysis, verification, and interface management. These capabilities should be practiced to a degree of detail not common today. The Mars Polar

Table 2.2. Systems and Example Application Areas

Types of Systems	Example Application Areas
Technological systems	Commercial aircraft and spacecraft
Product-centered infrastructure systems	Aircraft and spacecraft developers
Technological systems with human components	Aircraft-pilot systems; nuclear power plant – operator systems
Human-intensive systems	Hospitals, military units; commercial businesses
Process systems	Manufacturing processes, operational processes, maintenance processes
Socioecological systems	River-dam systems
Complex adaptive systems	Stock market and manufacturing systems
Systems of systems	Civil infrastructure systems of systems
Federations of systems	Jointly managed military systems or aviation control systems
Infrastructure systems	Individual civil infrastructure systems (fire, power, water, etc.)

Lander, which is discussed in Chapter 4, Case Histories, is an excellent example. If the strut vibration and sensor performance had been verified, it is unlikely that that accident would have ever occurred. This is also an example of the need for more detailed interface management to examine the negative interactions among system elements. It is an example of Woods' (2007) heuristic: "avoid hidden interactions."

The above principles are limited not only to the technological processes mentioned in Chapter 6 but also to the managerial processes. The peer reviews of Chapter 9, Governance, are an example of managerial processes that need to be performed rigorously. One objective of this process is to ensure that remote areas of expertise are considered. The Tacoma Narrows Bridge and the Comet aircraft, which are discussed in Chapter 4, Case Histories, are examples of remote areas of expertise whose neglect resulted in severe consequences.

To understand and implement system resilience, especially for technological systems, it is necessary to understand these systems at two levels. The first level is the technological system itself; this system is the hardware and software that perform the objective of the system. The technological system can be an aircraft, spacecraft, nuclear power plant, or chemical plant. To understand a system only at this level is to understand it superficially. An equally important level is the infrastructure system that designs, produces, operates, and maintains the system. Chapter 8, Resilience Architecting, will discuss this type of system.

To understand systems in the resilience context, it is necessary to define the system for each case study. For the Challenger and Columbia, for example, the system of interest is more than the flight hardware and software. The contracting agency, National Aeronautics and Space Administration (NASA), is a critical part of the system. In addition, the operational and support elements are integral parts of the system. The integrity of the flight equipment owed as much to these infrastructure elements as these elements did to the design of the flight vehicle itself. The Columbia Accident Investigation Board report (2003, p. 177), for example, states that the cultural factors were at least as important as the technical factors in that disaster.

2.2.2 Product-Centered Infrastructure Systems

So the question is, when a system, is either physical or organizational, will you find a "process" for implementing adaptability into the system? The answer is that it is highly unlikely. Following are a few steps to begin the process.

The first step is education. Key people in the organization need to know what adaptability is, why it is needed, and what its characteristics are. Although probably not many courses are available on adaptability, these need to be developed also. At the technical level, adaptability needs to be incorporated into design manuals and other guidance for design. All key design people need to understand adaptability.

The second step is to decide who has the responsibility, accountability, and authority (RAA) for implementing adaptability. One theme of this book has been that from a resilience point of view, there should be no dividing line between management and engineering. So, who is in charge, a manager or an engineer? The obvious answer is to put someone in charge who is responsible for both. That person may have a title like Chief Engineer. It does not really matter; what matters is that adaptability from both a managerial and engineering point of view is implemented.

In addition to the key person mentioned above, another person with RAA would be an adaptability expert who could advise the Chief Engineer on all decisions. Some organizations have a person called a Systems Architect, which is not to be confused with architects of houses and buildings. A System Architect would be familiar with the heuristics of Rechtin (1991 and 2000), Woods (2006b) principles, Billings' (1997) heuristics, and any other resources pertinent to adaptability. Although the Systems Architect may be known by any number of names, the function is critical to adaptability. In the design of any system, there are normally reviews of the design. This person would have a key role in these reviews to assure that adaptability is implemented. Chapter 6, Capabilities, discusses the concept of systems architecting and the roles of the systems architect.

O-rings and tiles are critical parts of space systems. Hence, even the smallest parts of a system, be they mechanical, electronic, or human, may be critical to the success of a system. An engineer, a component in the development system, who makes an error in a drawing, can cause a failure. No system requires zero defects; however, in all systems, a set of critical items is necessary for success. Sometimes these items can be very small. Chapter 10, Governance, discusses how to minimize errors in the development phase.

2.2.3 Technological Systems with Human Interfaces

A commercial aircraft, its pilot, and its maintainers are an example. The field that studies the interaction between the technological system and the human is called cognitive engineering. According to Norman (1982, pp. 378–382), "cognitive engineering is an interdisciplinary approach to the development of principles, methods, tools and techniques to guide the design of computerized systems intended to support human performance."

This type of system is important because it underscores the importance of the interaction between humans and the product system. Following are some heuristics of Billings (1997) on this subject.

In particular, Billings states that when the decision is made to give a task to a human or a machine, it is better to give it to the human if it can be done by the human. This statement recognizes the fact that humans are better judges of unusual conditions, such as those experienced when a disruption occurs. A case in point is the Nagoya accident discussed in Chapter 4, Case Histories. Although it is not clear that the pilot could have saved the aircraft involved, it is almost certain that the pilot understood that there was a conflict between him and the flight control system. However, the flight control system was unlikely to understand the intent of the pilot, which violated another Billings rule that every element of the system should understand the intent of the other elements.

The danger of the above heuristic is that it might be misinterpreted to imply the so-called "white scarf" syndrome, that is, the belief that humans are always better at any job. This term derives from the romantic image of pilots in the 1930s and 1940s wearing white scarves. This was not Billings' intent. Billings was only saying that for those situations in which humans could perform the task, they should do it.

Billings probably intended these rules to be applied mainly to aircraft systems. However, there is no reason to believe they would not apply to any cases in Chapter 4 in which both humans and software were involved. Chernobyl, as described by Chiles (2002, pp. 161–164 and 307), for example, comes to mind.

Although Billings may not have considered his work to be relevant to resilience or to adaptability, his heuristics seem particularly applicable to disruptions in which the two agents are humans and software, such as the Nagoya incident discussed previously. Although the Woods principles, above, seem valid and logical, there is no prohibition against suggesting other adaptability principles that will also add value. The Billings heuristics, as indicated below, are a worthy set to add to the set.

Billings' rules should not be construed to imply that the human is always right or that the human should always be in control. The last rule about "intent" implies that it is equally important for the software to understand the human and *vice versa*.

2.2.4 Human-Intensive Systems

Human-intensive systems can consist of any organization of any size. It can be a company, a government infrastructure, a hospital emergency room staff, a regional authority, or an entire society. This book pays particular attention to this type of system. Such a system is depicted later in this chapter under *The infrastructure system*. Human-intensive systems may contain hardware and software, but the predominant elements are humans.

As discussed below, a basic principle of the systems approach is that the boundaries of the system should be defined. In the case of Hurricane Katrina, the system is the complex set of government agencies that were responsible for dealing with hurricanes. The system also includes the levees and the U.S. Army Corps of Engineers who were responsible for designing and maintaining the levees. Westrum (2006b), for example, asserts that the root cause of the lack of resilience in the case of Katrina was a lack of leadership [by Federal Emergency Management Agency (FEMA)]. Others disagreed, saying that a truly resilient system would not depend on leadership.

It is not the purpose of this book to pose solutions to the failure of the Hurricane Katrina infrastructure. However, it is a classic example of a system of systems in which the component parts were disconnected. In a resilient system of systems, all the parts would work together to plan for transportation, resources, power, medical care, shelter, and a host of other factors. Whether this cooperation is facilitated by the government or the individual parts themselves is unimportant. The important factor is interelement collaboration.

2.2.5 Process Systems

Process systems consist of the tasks and actions performed by machines, humans, or any other elements of a physical system. Examples are manufacturing processes, operational processes, and maintenance processes.

2.2.6 Socioecological Systems

These systems are included because these systems are products of human intervention. For example, when humans build dams on rivers, the resulting system is a socioecological system. When humans insert species into a native population to avoid endangerment, a socioecological system is the result.

Liu et al. (2007) study six cases in which human systems interact with natural systems. This combination of systems can be referred to as a socioecological system. Liu et al. have observed that this interactionn is characterized by "nonlinear dynamics with thresholds, reciprocal feedback loops, time lags, *resilience*, heterogeneity, and surprises." As an example of resilience, the authors point to the action by humans in preserrving forests around the world through conservation and the addition of nutrients.

Guikema (2009, p. 1303) shows how integrating the design of buildings with the natural environment, such as mangrove swamps, "reduce the impacts of hurricane winds, hurricane surges, and tsunamis on infrastructure." In addition, Guikema suggests that these types of integration efforts can reduce the total life cycle costs on buildings by, for example, reducing heating costs and susceptibility to mold.

2.2.7 Complex Adaptive Systems

Both human-intensive systems and ecological systems are often called complex adaptive systems (CASs). By definition, CASs are adaptable; that is, they can adapt to recover from a disruption. Adaptability is discussed in Chapter 8, Resilience Architecting. According to Dooley (1996), "A CAS behaves and/or evolves according to three key principles: order is emergent as opposed to predetermined, the system's history is irreversible, and the system's future is often unpredictable." Many CAS examples are natural systems, such as a colony of ants. If there is no human intervention, then these CASs are out of the scope of this book except to the extent that we can learn from them how to be adaptive. For example, because a military aircraft squadron and a flock of geese have similar objectives, namely to fly in the same direction without colliding, it can be concluded that the logic is similar regardless of whether the aircraft is using humans or software flight control system to maintain its formation.

However, other examples of CASs are the stock market and manufacturing systems. Because these systems are human-intensive system and subject to architecting by humans (see below), they are in scope.

Many systems discussed in this book are, indeed, candidates to become CASs if they can achieve a level of resilience through adaptability. In Chapter 7, Infrastructure, and in the discussion below, infrastructure systems fall into this category. Pommerening (2007, p. 12) says that CAS theory "promotes a view of systems as being designed with an ability to adapt to change under conditions of uncertainty – in short, resilience." Chapter 8, Resilience Architecting, provides heuristics for achieving the state of resilience.

2.2.8 Systems of Systems

Another concept important to resilience is *systems of systems*. Systems of systems are collections of systems that need to work together to achieve a common goal. According to Sage (2008), in a system of systems each component system can function autonomously. A true system of systems has a limited degree of centralized control. However, when there is little centralized control, the system is called a federation of systems. Most infrastructure systems can be either systems of systems or federations of systems. For example, a police force, a fire department, a power system, and a water system can all work independently, but under the threat of a hurricane, for example, they should all cooperate to achieve resilience of the entire infrastructure in the face of the hurricane. For a federation of systems, integrated functionality can only be achieved through the effort of a central coordinating node and the cooperation of all the nodes, that is, component systems. The end result that is desired is the loose coupling described by Grote (2006). This loose coupling consists of continuous communication, agreement, and cooperation. As easy as this may sound, this is one of the more difficult goals of resilience to achieve.

Various commercial airlines acting together with an air traffic control system is an example of a system of systems. The more systems that interact with each other to perform a common purpose, the higher the likelihood that negative interactions will occur. These negative interactions are result in *disruptions*, which cause near misses and accidents. Chapter 3, Disruptions, discusses the two types of disruptions, each of which may require a different response: Type A disruptions—degradation of input—or Type B disruptions—degradation of systemic internal function, capability, or capacity. The most resilient systems avoid, survive, and recover from both types of disruption. This chapter discusses all three types of disruption and gives examples of each one.

Finally, it can be said that a system of systems has a strong potential for being a resilient system for two reasons: First, because by definition, each element of the system of systems can act independently, the system of systems will meet the criterion of "localized capability" discussed in Chapter 8, Resilience Architecting. This capability gives the system of systems its attribute of gradual degradation, which is another characteristic of a resilient system also discussed in Chapter 8. The second important feature of a system of systems to make it resilient is the attribute of interelement collaboration, which is also discussed in Chapter 8. This feature means that although the elements are independent, they must communicate and work together to help the entire system of systems achieve its functional purpose.

2.2.9 Infrastructure Systems

Another major system resilience enabler, which is shown in Figure 2.1, is the system infrastructure, discussed in Chapter 7, Infrastructure. The important principle to remember is that the infrastructure itself is both a system and a system of systems. The system resilience infrastructure is the set of all organizational entities involved in the success of a system. The key theme of this chapter is that an infrastructure should have flawless *interelement collaboration*, as shown on the figure. Interelement collaboration is one of the primary characteristics of adaptability as described in Chapter 8, Resilience Architecting. Interelement collaboration is a managerial function. However, it also applies to technological systems.

In the context of this book, the infrastructure is the largely human-intensive systems that develops, operates, and maintains the end product system: for example, an aircraft, spacecraft, or nuclear power plant. It is the larger enterprise system described above under the concept of system. In the cases of Hurricane Katrina and the New York Power Restortation both described in Chapter 4, Case Studies and Root Causes, the enterprise system is *the* system of interest. Such a human infrastructure system is depicted in Figure 2.2.

An infrastructure contains any organizational element that may affect system resilience. Hence, the scope of the infrastructure may vary from domain

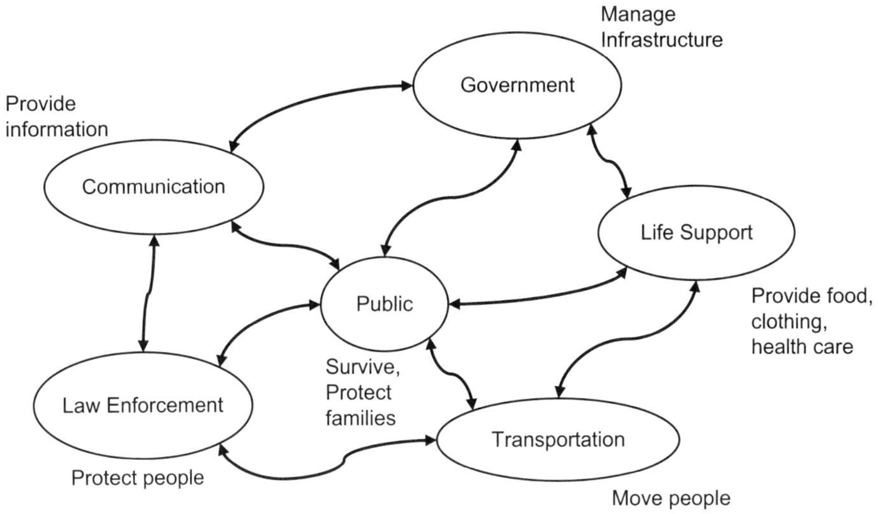

Figure 2.2. A Human-intensive infrastructure system.

to domain. Domains may include, for example, civil, military, commercial aircraft, and maritime infrastructures. However, in general, they will contain the developer, the maintenance organization, and the operators. It does not matter whether, for example, the operators and maintainers belong to the same organization, for example, an airline. The important thing is that they are separate nodes of the infrastructure system. Other nodes may include, for example, a test node or a regulatory node.

Figure 2.3 is a simplified conceptual view of a product-centered infrastructure system. Both Figures 2.2 and 2.3 are based on the Department of Defense Architectural Framework (DODAF) Operational View (OV-2). The point of these figures is that for resilience, all the elements of the infrastructure system should operate as a whole, not as a collection of parts. To achieve this goal, the cultural paradigm pertaining to organizational and contractual constraints discussed in Chapter 5, Culture, should be overcome.

For both technological systems and infrastructure systems, the point of the greatest potential weakness is at the interfaces between the elements. Rechtin (1991, pp. 29, 107) states that the "greatest leverage in systems architecting is at the interfaces." It can be concluded that the inverse of this heuristic is also true: namely, that the failure to consider interfaces will result in an inferior system.

The disconnects between the parts of the infrastructure became obvious in the case of the ValuJet aircraft that crashed because oxygen canisters had been improperly loaded into the cargo as described by Chiles (2002). In a traditional arrangement. the three segments of the infrastructure, the developer, the operator, and the maintenance company, are loosely connected by contractual agreements. This is not to say that contractual agreements are the source of

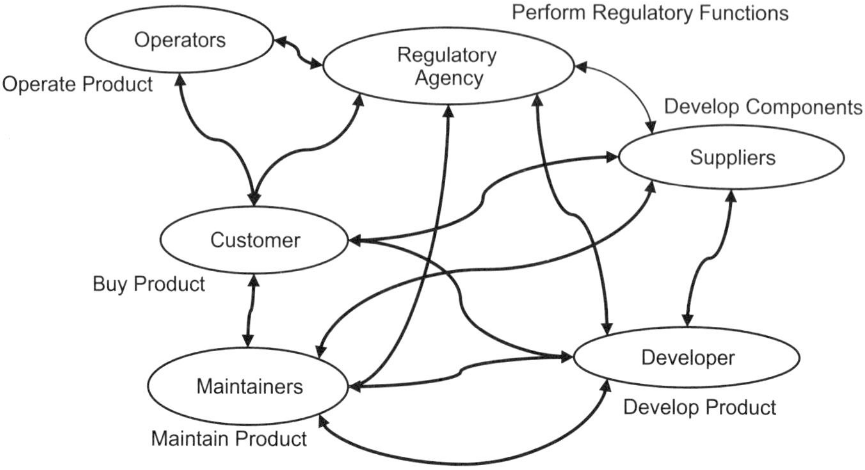

Figure 2.3. Operational view of a product-centered infrastructure system.

infrastructure weaknesses. However, it is to say that special attention should be given to the contracts that connect the various nodes in an infrastructure to make it resilient.

The idea that the infrastructure as a whole is not well understood. Tightening the bonds between the parts of this infrastructure is another challenge of resilience. This idea is a particular challenge in today's government and organizational structure in which organizational responsibilities are recognized primarily as local rather than global.

However, an infrastructure is more than a set of interconnected organizations as described in Chapter 7, Infrastructure. It also includes the capabilities of those organizations as described in Chapter 6, Capabilities. Of course, the culture of the people who populate the infrastructure is also a basic attribute of the infrastructure as described in Chapter 5, Culture.

The integrity of the organizational infrastructure responsible for resilience is paramount. The need to resilience puts a greater responsibility for the various nodes of the infrastructure to operate as a whole and not as a collection of organizations as described in the Infrastructure systems section in this chapter. This requirement also challenges traditional cultural paradigms as described in Chapter 5, Culture.

2.3 SYSTEM RESILIENCE CAPABILITIES

As shown on Figure 2.1, capabilities are divided into two broad categories, managerial and technical, both of which require methodologies beyond common practice in industry, government, and the world in general. The capabilities discussed here will be used as a benchmark for the case studies to be discussed in Chapter 4, Case Histories. They will be elaborated on in Chapter 6, Capabilities. In Chapter 8, Resilience Architecting, they will become the basis for the creation of resilience within systems.

There is no assumption that all the capabilities in this chapter need to be implemented on all systems in all domains. Chapter 12, Implementation, elaborates on this point. The degree of implementation will depend on the cost of the implementation, the consequence of a failure of the system the risks associated with that consequence, and the benefit-cost ratio, if determined, that would result from its implementation. The Federal Aviation Administration (FAA) has been most successful in determining the benefit-cost ratio, as discussed in Chapter 11, Cost.

Richard Feynman, a Caltech physicist (1988, p. 37), emphasizes his belief that there has to be a bond of common interest between management and engineering. It was this bond that made the Moon landings a success and this bond that had deteriorated on the Space Shuttle project. Feynman summarizes his theory as follows:

> So my theory is that the loss of common interest—between the engineers and scientists on the one hand and management on the other—is the cause of the deterioration in cooperation, which as you have seen, produced a calamity.

Managerial oversight capability is required to achieve system resilience. Understanding of resilience is needed at the highest levels of management. Management is responsible for all the other capabilities, both managerial and technical, are in place. Managerial oversight is required for every node of the resilience infrastructure as described in Chapter 7, Infrastructure, whether it is a human-intensive system or the managerial level of a technological system.

Especially for technological systems, a key managerial capability is contract management. Contract management is far more difficult in the resilience context than in the past because of the lack of quantitative criteria to put in contracts, as discussed below under *Contractual environment*. This difficulty results from the fact that traditional contracts depend for the most part on specifications that rely on verifiable requirements to design a product of desired performance. Resilience, however, relies on heuristics, which are not verifiable. The most likely path to this change lies in process-based procurement rather than requirements-based contracts. In these cases, contracts would direct the developer to use resilience design attributes and submit the results to the procurement agency for concurrence. This method is used today, but in the future it may be more common as resilience becomes a higher priority.

Cross-scale connectivity is the basic capability, defined by Woods (2006b), which is the basic set of linkages among the nodes of the system, again whether human or technological. Cross-scale connectivity involves communication, agreement, collaboration, data sharing, and resource sharing among the organizations. Cross-scale connectivity is essential for the various nodes of a multiorganization system of systems, such as an urban infrastructure. It is also the linkage among the various nodes of a single organization, including suppliers and production. It links elements of the system across the life cycle. For example, it links development with operations and support nodes. Particularly important is the cross-scale connectivity with regulatory agencies. Cross-scale connectivity leads to the attribute of interelement collaboration used in this book. Interelement collaboration is more than interaction; it is the conscious cooperation, coordination and collaboration, among the elements.

Another managerial capability is system resilience governance. The essence of independent reviews is that personnel outside the program provide comments and insight into potential flaws and risks. Similarly, peer reviews lean on experts in various disciplines to do the same. These steps will tend to reduce the likelihood of the types of accidents discussed in Chapter 4, Case Histories. Corrective action is a key process designed to mitigate the "drift' toward brittleness described by Woods (2007). Managerial oversight is required to establish decision-making philosophy in both the system design and the system operation.

Other capabilities, such as the more advanced treatment of risk, are also covered as part of managerial capabilities. These include the consideration of the multiple low-probability risks discussed by Epstein (2006) and the dynamic, nonlinear risk analysis of Leveson et al. (2006). Among risks to be addressed are those associated with technical maturity, cost, and schedule factors.

The final managerial responsibility is culture. Abundant evidence, for example, Vaughn (1996, p. 406) and the Columbia Accident Investigation Board report (2003, p. 177), suggests that risk-inducing cultural paradigms, or mindsets, are responsible for the lack of success in many systems. Chapter 5, Culture, lists many of the better-known risk-inducing paradigms and suggests ways to convert them into beneficial paradigms.

Technical capabilities also represent a new path both for technological and human-intensive systems. Most systems in the past have been created using the analytic methods described below under section 2.5, *Systems approach*. These methods rely heavily on designing a system that meets verifiable requirements. As described by Woods (2007), these methods are generally not adequate to realize resilient systems. An essential technical capability is holistic analysis. Holistic analysis, on the one hand, allows the analyst to take into account the many complex factors that are just not tractable by analytic methods. Holistic analysis takes into account the relationship among the systems elements. Analytic analysis, on the other hand, focuses on the capabilities of individual components.

The creation of *adaptability* to handle disruptions is, for example, not treatable by analytic methods, as discussed in Chapter 8, Resilience Architecting. Holistic analyses include, for example, the use of heuristics, rules, laws, and attributes. According to Woods (2006b), five principles constitute adaptability: First, there is the buffering capability that a system can absorb a disruption without a total loss of capability. Second, there is the flexibility to restructure itself. Third, there is the margin to that a system can continue to function relative to the boundary of its performance envelope. Fourth, there is the tolerance to graceful degradation in the face of imminent collapse. Finally, there is the interelement collaboration, discussed above, so that elements of the system are acting in a mutually beneficial fashion.

This is not to say that analytic methods should be completely abandoned. Case studies have shown that analytic methods might have helped but that they should have been performed in more detail and over a wider range of conditions. For example, analytic methods might have been beneficial to the Mars Polar Lander if more detailed requirements analyses of the struts and software had been performed as described by Leveson (2002b, p. 268). In the same case, detailed interface management would have identified the negative interaction between the elements. Improved maintenance requirements across organization boundaries might have prevented the ValuJet incident described in Chapter 4, Case Studies and Root Causes.

Another discipline that departs from the past is safety. For example, Leveson et al. (2006) and Jackson and Hann (2004) show how focusing just on the design is inadequate and how organizational factors need to be folded into the safety analysis.

2.4 CULTURE

Another major element of the system resilience concept diagram, as shown on Figure 2.1, is culture. The Columbia Accident Investigation Board report (2003, p. 177) found that cultural factors were at least as important a contributor to the Columbia disaster as technical factors. Vaughn (1996, p. 406) found the NASA culture of accepting risk in the pre-Challenger period to be a normative, or established, norm. Hence, risk denial can be said to be a negative cultural factor. Likewise, positive cultural factors can be identified. Deference to expertise is an example of a positive cultural factor.

Of course, there is the famous Titanic Effect as mentioned by Leveson (1995, p. 57). This is the irrational belief that a system is safe.

The term culture, as discussed in Chapter 5, Culture, may refer to the individual, organizational, or national beliefs that govern our actions. Although the cultural influences on resilience are well documented by Vaughn (1996, p. 406) and the Columbia Accident Investigation Board report (2003, p. 177), this aspect has received little attention in industry or government for several basic reasons. First, culture is not technical or quantifiable. Second, when it is considered, the solutions are simplistic and not based on any sound analysis or research.

The most common approaches are training and speeches by executives. These approaches have been shown to be singularly ineffective. Other popular approaches include creating teams and established processes. The jury is still out on these approaches also. Organizational psychologists favor methods of self-discovery, such as communities of practice described by Wegner (1998). In any case, the silver bullet has not been found. Hence, much more research is needed in this area. Irrespective of the approach, in view of the findings, such as those of Vaughn (1996, p. 406) and the Columbia Accident Investigation Board report (2003, p. 177), it can be concluded that this factor cannot be ignored.

2.5 DISRUPTIONS

Another important element on the system resilience concept diagram of Figure 2.1 is disruptions. Disruptions are the events that jeopardize the functionality of the system, especially if the system lacks resilience or, better put, is brittle. Disruptions are described at length in Chapter 3, Disruptions.

For now, it is sufficient to understand how disruptions occur. Once we understand the nature of disruptions, we can then begin to observe how the attributes, laws, and constraints can be brought to bear to create systems that are resilient to any type of disruption.

The two types of disruptions defined in Chapter 3, Disruptions, are either internal or external disruptions. Whether disruption is internal or external is dependent on how the system is defined.

Any new phenomenon, at least to the system designer, can be considered an external disruption, falling into Type A, disruptions of input. For example, before 1958 the presence of the Van Allen radiation belts was unknown. Similarly, the existence of turbulent aerodynamic layers on the Tacoma Narrows bridge in 1940 was unknown.

Second, if a pilot of an aircraft makes an error, is this an internal or external disruption? Using the definition of a system from Chapter 3, it is more accurate to say that the pilot is part of the system, so that any error that he or she might make would be considered an internal disruption. Hence, pilot errors would be listed in the Type B disruptions, or degradation in function, capability, or capacity. Two-agent disruptions, as in the cases of the Mars Polar Lander and the Nagoya accidents, which are described below, are also internal disruptions.

The Mars Polar Lander, discussed in Chapter 4, Case Histories, is one example of two-agent disruption. In this case, the software interpreted the Lander strut vibration to be a signal of the vehicle's having landed. The software shut down the engines, and the vehicle crashed into the planet.

Because modern systems are more software intensive and involve more interaction between software and humans, they will be more vulnerable to disruptions than in the past. Space shuttles and commercial aircraft contain increasingly large numbers of software modules. When the software and humans act against each other, as in the case of the Nagoya accident, the results can be catastrophic. This accident is described by Leveson (2002, pp. 30, 47, 64, 147, 249).

As discussed below, many systems are really systems of systems, that is, several systems working together to achieve a common goal, as defined by Sage (2008). We saw before that the Hurricane Katrina infrastructure was a system of systems. In short, systems of systems are particularly subject to network disruptions. That is to say, when one element of a system fails, this failure can propagate into the elements of the other systems of which the system of systems consists.

2.6 SYSTEMS APPROACH

In Figure 2.1, there are two elements called *holistic methods* and *analytic methods*. The systems approach is a general term used for the organization and management of complex systems and is the foundation for both analytic and holistic methods.

Although many sources define the systems approach, the definitions of TSI (2009) and the University of Washington (2009) are typical. Table 2.3 reflects elements of both sources. The INCOSE Fellows (open discussion, February 3, 2009) have reviewed and concurred on the list in Table 2.3. With regard to this table, first, the identification of system elements facilitates analysis by enabling each element to be treated according to the disciplines required for that element. Second, subdivision of elements into subelements allows concentration on each element and the parts of the element. Grouping of elements facilitates analysis when a group of elements contributes to a common goal or function. Defining the boundary of a system helps identify what elements contribute to the goal of the entire system. Identification of the function of each element enables an analysis of how each function is to be performed. An analysis of the interactions among the elements enables the designer to determine how the elements perform together as a whole to achieve the goal

Table 2.3. The Systems Approach

- Identification of system elements
- Subdivision of elements into smaller elements
- Grouping of elements
- Identification of system boundary
- Identification of the function of each element
- Analysis of the interactions among elements
- Identification of the system environment
- Identification of emergent characteristics of the system
- Synthesis of the system
- Verification and validation of the system

of the entire system. Identification of the environment of each element allows the analysis of constraints on the performance of each element. Because the environment of each element may be different, there will be different constraints on each element. Finally, the elements, when acting together, exhibit *emergent* properties. These properties are the characteristics of the system as whole that are not possessed by the individual elements. This latter characteristic is the one that may be considered most essential because it is the one that defines the true function of the system as a whole. Checkland (1999, p. 314) defines emergence as follows:

> The principle that whole systems exhibit properties which are meaningful only when attributed to the system as a whole, not to its parts—e.g., the smell of ammonia. Every model of a human activity system exhibits properties as a whole entity which derive from its component activities and their structure, but cannot be reduced to them.

Thus, resilience itself is an emergent property that cannot be derived from its components, only from the entire system.

These attributes can be applied to any kind of system, including technological systems (hardware and software) and human-intensives (such as organizations). As can be observed above, the systems approach is a set of high-level attributes for the analysis of systems. Two types of methods have emerged as ways to implement this approach. They are called the holistic methods and the analytic methods. Although most sources, for example, the University of Notre Dame described by Wilbur (n.d.) and the INCOSE Handbook (2006), describe these methods as two distinct methods, it is difficult to design a system without using a combination of the two. Even the INCOSE Handbook discusses holistic methods to some extent.

According to Wilbur (n.d.), holistic methods "are essentially couched in the belief that the whole is not only greater than the sum of the parts, but that the parts are related in such a way that their functioning is conditioned by their relationship to each other." This methodology is mentioned first because it is more appropriate for the study of resilience, especially for more complex systems, especially systems with human components. One primary technique that employs the holistic approach is *systems architecting*, to be discussed in Chapter 6, Capabilities. Systems architecting relies heavily on the use of heuristics, attributes, rules, and laws. Heuristics are the lessons from experience that allow the designer to formulate the architecture of a system. Some of these heuristics are discussed in Chapter 8, Resilience Architecting. Systems architecting employs both holistic and analytic methods, but it relies strongly on holistic methods.

Because holistic methods include any consideration that treat the system as a whole and not just individual elements, the consideration of culture in Chapter 5, Culture, and risk in Chapter 6, Capabilities, could be holistic considerations. The various environments discussed at the end of this chapter are also holistic considerations.

The second type is called analytic methods. These methods consist primarily of defining a set of well-defined requirements, creating a design from those requirements, and verifying the final solution. This methodology is used primarily for technological systems. Woods (2007) has pointed out that the analytic methodology alone generally does not result in resilient systems. However, to a limited extent, the analytic approach is useful for defining detailed aspects of a system that may be used as a starting point.

Nadler and Chandon (2004, p. 7), in their comparison of holistic with reductionist, or analytic, approaches, assert that the reductionist approach is based on Cartesian thinking, in which every system can be broken down into its component parts, and each part can be treated independently. The holistic approach considers the interaction among the parts. The holistic approach is particularly important when studying disruptions that occur because of the interaction between the components, even when each component operates as designed, as discussed in Chapter 3, Disruptions.

In short, with a product system, such as a commercial aircraft or nuclear power plant, the analytic approach can be employed, in part. However, to account fully for the unpredictable aspects of humans and software, advanced approaches, such as systems architecting, are required. At the present state of the art, the decision whether to employ holistic or analytic methods is not well defined. All that can be said is that both are required for resilience.

2.7 SAFETY

Figure 2.1 also shows *advanced safety methods* as an element of the system resilience concept diagram. *Safety* is another term whose meaning needs to be viewed in a broader context to understand system resilience. Thus, one could say that safety, which is achieving a system goal or objective without loss of life, is the whole point of system resilience. In S. Jackson (2002), safety is discussed in three contexts (Figure 2.4).

First, industrial safety is the protection of the work force against potential accidents. Industrial safety concentrates on such incidents as slipping on a wet floor or being injured by a piece of machinery.

Second, system safety focuses to a great degree on the product system itself. It asks the question: Given that the system is properly maintained and operated, is it likely to fail with disastrous consequences? However, in many organizations, system safety is not concerned with organizational factors. System safety is primarily concerned with the design of the product system. Reason (1997) concentrates on the organizational factors. As shown in Chapter 4, Case Histories, major accidents are rarely caused by the realization of an identified safety hazard. On the contrary, the root causes of major accidents are normally beyond system safety considerations.

Another characteristic of system safety is that it typically concerns defined failures, whereas the goal of resilience is to create systems that can avoid accidents and survive and recover from unknown and unprecedented disruptions.

Figure 2.4. Three types of safety.

In addition, system safety does not normally concern itself with the *anticipation* factor in system resilience. Leveson's chapter called Fundamentals of System Safety (1995, pp. 145–168) does not mention anticipation as an aspect of system safety. That is to say, a characteristic of resilience is the ability to detect an imminent threat to the system and provide the means and the opportunity to avoid any accident that might result from the threat.

Finally, S. Jackson (2002) defines a third type of safety, which is organizational safety. Organizational safety is the resilience of the organization to contributing to failure and loss of life by the product system. This is the type of safety of interest to Reason (1997). Figure 2.4 illustrates the relationship of industrial safety, system safety, and organizational safety. The point is that the three types of safety are not mutually exclusive. There is a strong interaction among all three.

Woods (2006, p. 320) makes the following point:

> Safety organizations must be involved in enough everyday organizational activities to have a finger on the pulse of the organization and to be seen as a constructive participant in the organization's activities and decisions that affect the balance across safety and production goals.

Whereas this is an excellent point, most organizations, either governmental or private, may find this principle at odds with current practice. In theory, safety organizations have some interest in organizational issues; in practice, this aspect is limited. For example, MIL-STD-882 (1993) calls for the establishment of a safety organization. However, the focus of this standard is on product safety rather than on organizational aspects.

Leveson (1995) has the most extensive discussion of safety and its role in resilience engineering. Leveson strongly emphasizes two points: first, the global view of safety, in line with Woods' comments, above, and second the top-down approach to safety using the systems approach. Hence, Leveson's view of safety is in total agreement with the goals of system resilience.

Hence, all the above considerations constitute *advanced* safety in the context of this book. None of these considerations is precluded from safety in current analysis, but in practice, the scope of safety analysis has boundaries. An expansion of these boundaries will enhance resilience.

2.8 ENVIRONMENT

For product systems, especially technological products such as aircraft and space vehicles, the term *environment* normally refers to the physical environment with such parameters as temperature, pressure, vibration, and so forth. For human infrastructure systems, the term takes on a broader meaning. ISO/IEC 15288 (2002) stipulates that the definition of the environment is fundamental to the system definition. The following paragraphs outline some principal environments critical to system resilience.

2.8.1 The Cultural Environment

Culture is what a society, either an organization or an entire nation, believes to be true. For a system to be resilient, this set of beliefs should support the goal of resilience. Sometimes, the cultural environment is detrimental to system resilience. Chapter 5, Culture, outlines both the cultural characteristics that reinforce system resilience and those that are detrimental. This chapter also outlines various approaches for either changing the cultural environment or working within it.

2.8.2 The Economic Environment

There is no system that does not exist within an economic framework. It is the human infrastructure system that determines the economic resources and constraints. A common assumption is that financial considerations are managerial as opposed to technical in their import. On the contrary, financial resources determine the extent of technological advances possible. At the same time, financial constraints create unacceptably high technical risks, as described in Chapter 6, Capabilities. Of course, there are also cost and schedule risks, but technical risks are most pertinent to system resilience.

2.8.3 The Regulatory Environment

Each domain has its own regulatory environment. In the United States, commercial aviation must abide by FAA regulations. Other industries, such as nuclear, chemical, and railways, have their own regulatory agencies. Most regulations focus on safety; however, other factors, such as noise and air pollution, are also regulated. Some major accidents have been attributed to lax enforcement of regulations, for example, the 1947 Texas City disaster described by Hughes (1997). Reason (1997, pp. 160–161) cites the King's Cross Underground disaster as a case in which the regulator and the operator (London Underground) had too close a relationship, as described in Chapter 4, Case Histories.

2.8.4 Contractual Environment

Almost all large systems exist within a maze of contractual agreements. Developers have contracts with customers and with suppliers. Sometimes these contracts have rigid performance and safety clauses. Other contracts require only the delivery of a product with little said about quality. If the product fails, the developer or supplier has no responsibility because the customer or developer has accepted the product. The writing of a good contract is essential to system resilience. However, such a contract is inherently difficult. It is easier to write a contract for a technological product with verifiable requirements. Writing a contract for a resilient system is much more difficult.

2.8.5 Political Environment

Whereas it can be said that all systems exist within a political environment, the extent to which this environment influences resilience varies. The most notable example is Challenger, discussed in Chapter 4, Case Histories. For example, Vaughn (1996, p. 372) states that "NASA's environment of competition and scarcity made political accountability a part of NASA culture in the 1970s." Although the Columbia Accident Investigation Board report (2003) does not specifically mention political factors in its conclusions, the presence of several Board members with specialties in politically sensitive programs attests to the political climate of Columbia.

The political environment has a direct impact on technical, cost, and schedule risks as discussed in the *Risk* section of Chapter 6, Capabilities. It is also responsible for the genesis of the Conflicting Priorities paradigm discussed in Chapter 5, Culture.

Where we observe the most important influence of the political environment is in the consideration of large infrastructure systems of systems, such as the Hurricane Katrina and Metrolink 111 accidents discussed in Chapter 4, Case Histories, and in Appendix B, respectively. What we observe in both cases is that the system of interest is a large system of systems in which there is little central control. These cases raise the question of how resilience is maintained under such circumstances.

2.8.6 Organizational Environment

Organizational environment can have different characteristics and effects. Some of these are strongly related to the cultural environment as discussed in Chapter 5, Culture. For example, the organization may be vertical or flat. Vertical organizations tend to have many layers of authority, and communications between layers is difficult. Flat organizations attempt to reduce these layers with the goal of improving communications. However, there is a limit to the degree of flatness itself. If an organization is too flat, then communications may suffer also.

Organizations may be either authoritarian or permissive. Authoritarian organizations prohibit any action without specific permission. This is not to say that either authoritarian or permissive organizations are more conducive to the implementation of resilience attributes. It is simply the environment in which implementation occurs. What is more important is to what extent management encourages self-discovery through such methods as communities of practice as described in Chapter 5, Culture.

Organizations may have many boundaries and contractual constraints, as discussed in Chapter 5, Culture. Some methods for dealing with organizational constraints from a resilience point of view are discussed in Chapter 7, Infrastructure.

2.8.7 Geopolitical Environment

The term *geopolitical environment* encompasses all of the factors that differentiate one country from another or, in a broader sense, the first world from the third world. Included are economic systems, political systems, geographic environment, and infrastructure systems.

Most incidents discussed in this book are first-world catastrophes. The advantages of a highly developed infrastructure in first world countries render the fatalities small compared with catastrophes in the third world. In an interview between journalist Dreifus (2006) and geophysicist John C. Mutter, Mutter compares a 2005 earthquake in Pakistan that killed an estimated 100,000 people with the Northridge [California] earthquake of 1994 in which only 63 people died. Both earthquakes were of similar magnitudes. Mutter explains, "they [richer countries] have all kinds of buffers—building codes, advance warning systems, insurance, first responders." Mutter goes on to clarify that this disparity also exists within the United States. He points out that most fatalities during the Hurricane Katrina disaster were poor and elderly.

All this raises the question: Do the principles of this book apply equally well to third-world countries as to first-world countries? The answer is yes, but the resources, and the buffers, will be vastly different between the two.

Although help for the third world is far away, some steps are being taken. For example, a worldwide association of organizations and societies, called Global Earth-Observation System of Systems (GEOSS, 2006) is developing a system of information regarding earthquakes, tsunamis, hurricanes, and other destructive phenomena. This system will fulfill the attribute of interelement collaboration as discussed in Chapter 8, Resilience Architecting. The question of resources is a longer term issue.

2.8.8 Physical Environment

Last but not least is the physical environment. For most product systems, for example, commercial aircraft and space systems, this is the primary environment of interest. Even this environment can experience wide ranges of severity in, for example, temperature, wind velocity, humidity, and other parameters. For an aircraft designed to operate in both Alaska and Saudi Arabia, the range of environments is large.

But how about environments whose extremes are experienced only infrequently or whose extremes are completely unknown: The task then is to determine what is the extreme with very low probability. For example, should the dyke be built for the 100-year flood or the 1000-year flood?

One could say that the two aircraft that flew into the twin towers were an "unknown environment." This is an example of a Type A, or input disruption, discussed in Chapter 3, Disruptions.

2.9 MISSION ASSURANCE

Another term in current use is *mission assurance*. This term is used mainly in the space and military domains where the product system is said to perform functions in support of a mission. Because NASA oversees space system development in the United States, its mission assurance approach is typical as described in its Process based mission assurance (PBMA) knowledge management system (2008). It is awkward, however, to say that a nuclear power plant, a chemical plant, a hospital, or a hurricane infrastructure has a mission. They have goals and objectives, but mission is not a normally used word.

The usefulness of this term is that it has to do with all the processes and disciplines that are necessary, for the most part, to achieve the accident avoidance aspect of system resilience in those domains, but it does not always address the survival or recovery aspects except, in some cases, military scenarios. Furthermore, mission assurance normally focuses on the rigorous application of processes, which are usually based on analytic methods, rather than holistic methods. For example, the NASA plan for safety, reliability, maintainability, and quality demonstrates an emphasis on known hazards, requirements, and testing as reflected in its *Safety, Reliability, Maintainability and Quality Provisions for the Space Shuttle Program* (1997).

2.10 FURTHER EXPLORATION

1. Examine some case studies in Chapter 4, Case Histories, and summarize the environments for each one. These include the economic, cultural, political, regulatory, legal and geopolitical environments.
2. Study publicly available literature on mission assurance and summarize the differences between mission assurance and resilience.
3. Study some of the publicly available articles on complexity science and summarize how it fits into the study of system resilience.
4. Study some of the publicly available literature on the systems approach and show how the holistic and analytic methods fit into it.
5. Examine some of the literature on enterprise systems within the literature and explain how it relates to the infrastructure system within the study of system resilience.
6. Examine recent literature on the subject of system safety and explain how an expanded scope fits into the subject of system resilience.

Chapter 3

Disruptions

"Design the elements to make their performance as insensitive to unknown or uncontrollable external influences as possible."

—Eberhardt Rechtin (1991)

Accidents are the result of disruptions, and the resilience of a system to disruptions will depend on the nature of the disruption. The goal of resilience is to avoid, survive, and recover from disruptions.

This chapter describes two types of disruptions, Type A and Type B, and the implications of disruption type on resilience. Type A disruptions consist of disruptions from outside the system that cause a loss of function of the system. Type A disruptions also result from a change of environment. Type B disruptions are systemically realized. They are the loss of function, capability, or capacity that results from systemic failures. For each example, the principal implication will be discussed, and a more comprehensive discussion of how to create resilience will be discussed in Chapter 8, Resilience Architecting.

Of course, whether a disruption is external or systemic depends on the boundaries of the system. A principle of the systems approach, which is discussed in Chapter 2, System Resilience and Related Concepts, is that the boundary of the system should be defined. As explained in Chapter 2, the boundary of a system may encompass many elements and therefore may be very large.

Next, the study of disruptions only addresses one phase of resilience, namely, the avoidance phase. So it could be argued that if a disruption can be avoided, then survival and recovery would not be of interest. However, the study of

Architecting Resilient Systems: Accident Avoidance and Survival and Recovery from Disruptions.
By Scott Jackson

disruptions is important from two points of view: First, if disruptions can be predicted and avoided, then major catastrophes can also be avoided. However, it has been observed that many disruptions simply cannot be avoided, either because they emanate from unpredicted sources, or because they result from highly random sources, such as human error. Second, by studying disruptions, it is possible to understand better how to treat survival and recovery.

Finally, the question must be asked: Does the study of resilience address probabilities that are smaller than are traditionally dealt with in practice? There are several answers to this question.

First, this is not to say that disruptions are not an important consideration in resilience, but resilience is broader than disruptions and their probabilities. Although minimizing disruptions is a major part of avoiding an accident, surviving and recovering from the disruption are also important parts of resilience.

Next, we must ask whether the probabilities are as small as they seem to be. The observation of Feynman (1988) is that the likelihood of a catastrophic accident may be considerably larger than generally assumed. Feynman did not determine what these other factors are, but he did determine that there was a considerable difference in opinion between managers and engineers about what the probabilities are.

So yes, in summary, resilience looks for failures beyond conventional methodologies, and second, resilience attempts to structure an architecture that will survive and recover from a disruption regardless of the probability.

3.1 TYPE A—DISRUPTIONS OF INPUT

Disruptions of input are an externally caused loss of function. These disruptions can result from random external phenomena. For example, an aircraft flying close to another aircraft can affect the second aircraft with its aerodynamic wake. If the wake is strong enough, it could affect the structural integrity of the second aircraft. The random direction and nature of the wake could affect the software of the second aircraft and lead it to misinterpret the angles of attack and yaw of the second aircraft.

Because disruptions of input, by definition, are caused by outside intervention, the primary need of systems sensitive to these effects is the detection of these effects and the ability to perform corrective action, regardless of the effect. Another implication is that the system needs to be tolerant to outside disruptions; that is, it should fail gradually rather than abruptly.

The Concorde supersonic aircraft disaster in the year 2000 described in the Bureau Enquétes Accidents (BEA) (2000) report is an example. The aircraft failed suddenly and catastrophically from a small strip of metal on the runway. There was no detection, no possibility of any corrective action, and no gradual degradation.

When the size of a disturbance is unpredicted, or unpredictable, resilience is most difficult. Is it possible to make a system resilient to a disturbance the *size* of which is unknown? One can collect data on the history of certain phenomena, for example, tides, winds, and temperatures. But what if a system encounters a disturbance greater than any historical value? Or perhaps, if a disturbance is completely unexpected, like the twin towers attack, then whatever magnitude the towers were designed to becomes irrelevant.

The primary implication for disruptions of environment is that systems need to be built for the worst-case environment plus an adequate margin for uncertainty. For commercial aircraft, this margin is normally 50%. Of course, this conclusion pertains to all elements of the system: water, power, transportation, and so on.

The most obvious examples of disruptions of environment are natural disasters, such as hurricanes, tsunamis, tornados, earthquakes, floods, fires, and droughts. Westrum (2006b), for example, discusses the failures of government agencies to deal with the Katrina hurricane of 2005. A far worse disaster, however, was the Galveston hurricane of 1900, described by Larson (2000), in which over 6000 people perished.

The predominant resilience aspect for dealing with this type of disruption is Woods' principle of cross-scale interaction (2006d). Simply put, cross-scale interaction is both communication and collaboration among all the elements of the civil infrastructure system charged with responding to natural disasters. These elements include power, water, law enforcement, medical needs, and transportation. Chapter 8, Resilience Architecting, uses the term *interelement collaboration* as one of the attributes of a resilient system.

3.2 TYPE B—SYSTEMIC DISRUPTIONS OF FUNCTION, CAPABILITY OR CAPACITY

The disruption of function, capability, or capacity is most evident in technological systems. It happens when a component fails, for example, in the case of Challenger. These types of failures can be categorized as reliability or safety failures, which can be treated by traditional analytic methods. However, these types of failures are less common in recent years. More recent sources, for example, Vaughn (1996), have pointed to a lack of attention to risk that may have been the true root cause of the disturbance.

If a software error, for example, sends an erroneous signal to another system node, then this is an example of a loss of function disruption. One of the best examples, related by Eyles (2004), is the guidance software failure on Apollo 11 that nearly resulted in a landing failure on the moon. Fortunately, the humans as backups managed to save the day showing that the system was resilient.

According to Aviation Week and Space Technology (2007, p. 23), an interesting, but benign, case of Type B disruptions occurred on February 19, 2007 when a squadron of F-22 Raptors lost the use of their navigational computers because the software had not been programmed, or tested, to account for crossing the International Date Line. Fortunately, the pilots were able to fly the aircraft manually and return them to their base. Despite their safe return, this is an illustration of the small, but avoidable, mistake that could have had tragic consequences.

Another example of Type B disruptions is the Nagoya accident described by Zarboutis and Wright (2006). In this accident, the pilot, a human component, innocently provided an incorrect command to the flight control system, a software component. Thus, a human component can be in conflict with a software component. In this case, no one won; the pilot, the aircraft, and all the passengers lost. In the Nagoya case, it is not useful to blame the pilot and end the discussion there. What is important is that the entire system, pilot and aircraft, were not adaptable, as described later. There are many ways to increase the adaptability of the pilot-aircraft system, either to give the pilot more options or to let the flight control system adjust to the situation. Future designers will understand these options.

3.2.1 Component versus System Failures

Leveson (2008) divides accidents into two main categories, those that result from component failure and "system" accidents. Component failures are discussed below as disruptions of unreliability. Leveson notes that these failures are normally characterized as random. However, this type of disruption does not preclude the possibility that the root cause of the disruption was managerial in nature. That is to say, the reliability of the component in question may have been inadequately analyzed and predicted.

Leveson (2008) lists four types of system accidents. The first is interaction among components. This type of disruption is discussed below in the category of multiple-agent disruptions. Once again, this type of disruption can be traced to either a system that is so complex that it would be virtually impossible to predict the interaction or a managerial failure to review and predict this interaction. The latter example may have been the case for the Mars Polar Lander discussed in Chapter 4, Case Histories. Leveson notes that the introduction of the computer has made it more difficult to predict these interactions because the software programmer must predict all possible scenarios in which the code must perform. Another factor cited by Leveson is the introduction of new technology that also exacerbates the interaction effect.

3.2.2 Disruption of Unreliability

When a component fails in a system, the question is as follows: Did it fail because it reached the end of its life prematurely or because of some other cause? If it reached the end of its life prematurely, then it can be called a disruption of unreliability, another Type B disruption. A disruption of unreliability occurs when a component has a mean time between failure (MTBF) of a certain amount and the failure occurs on the near end of its failure distribution curve.

This type of disruption assumes that the MTBF has been properly verified; that is, the MTBF has been based either on historical data or on test data. If the component has not been properly verified, then the disruption can fall into the category of management failure to assure proper verification.

Both Apollo 13 and Columbia are examples of disruptions of unreliability. For Apollo 13, the failure was of the power system. For Columbia, the failure was of the tiles on the surface of the vehicle. Chapter 4, Case Histories, shows that Apollo 13 was a resilient system; that is, it was able to recover from the disruption, whereas Columbia was not resilient; that is, it did not recover.

Challenger, however, was not a disruption of unreliability. It can be called a verification failure, that is, the failure, to verify the component, the O-rings, within its operating envelope. Similarly, the Alaskan Airlines Jack Screw failure can be called a maintenance failure rather than an unreliability failure. All of these cases are discussed in Chapter 4, Case Histories.

3.2.3 Disruptions Caused by Latent Conditions

Reason (1997, pp. 10, 11) distinguishes between active failures and latent conditions. Active failures are immediate and short lived. Active failures, on the one hand, happen on the "sharp end" of the system operation. They may be the result of an error on the part of an operator, such as the pilot in the Nagoya accident. Latent conditions, on the other hand, are long-standing deficiencies in organizational policies, procedures, training, and other factors. The failure to anticipate the interaction between the struts and the software on the Mars Polar Lander is one example. In short, latent conditions are organizational in nature. Chapter 7, Infrastructure, elaborates on some of the organizational attributes that enhance resilience.

3.3 AGENTS AND THEIR ROLES IN DISRUPTIONS

An agent is any element of a system. In this chapter, we shall refer to hardware, software, and human agents. Agents can also be external to a system. Hardware refers to structure and to mechanical, electrical, hydraulic, and devices other than electronics. Each type of agent, because of its inherent capabilities, contributes in varying degrees to disruptions. The following paragraphs discuss these agents in descending order of importance to disruptions.

3.3.1 Human Agents

Human agents are a two-edged sword. On the one hand, they are capable of the most unpredictable actions. The human pilot in the Nagoya accident was the central agent in the events leading up to the accident. On the other hand, humans are the most adaptable agents capable for recovery from any given disruption. The Apollo 13 mission, for example, was saved by the quick and ingenious actions of the crew. In this case, which is a Type B disruption, described by Leveson (1995, pp. 558–563), the loss of function was a power loss in the main module. The crew was able to take advantage of the resilient attribute of *flexibility* by moving to the smaller landing module. Chapter 8, Resilience Architecting, provides a deeper discussion of flexibility.

Many systems are almost completely human. Enterprises and organizations are examples. The infrastructure responsible for handling hurricanes, such as Katrina, is a human-intensive systems.

It cannot be denied, however, that beneficial results may emerge from accidental situations that were completely unexpected, either as a case of weak disruptions or strong disruptions. Take Alexander Fleming's discovery of penicillin, for example, described by Bellis (2009). Fleming had absolutely no idea that such a discovery would result from a routine laboratory experiment.

From a system point of view, it is reasonable to conclude that the reason that the air traffic control system is not totally automated is that it is assumed that humans are much more capable of detecting and handling unpredicted situations.

There is one lesson to learn from these examples: Almost all positive outcomes result when a human agent is involved.

3.3.2 Software Agents

Software agents also contribute to disruptions. Software was one of the contributing agents in both the Nagoya and Mars Polar Lander accidents. The performance of software is dependent on the ability of the software designer to anticipate all the situations in which the software has to operate. If a situation occurs that has not been anticipated, then the software will continue to execute the algorithms it was designed to perform whether it was supposed to or not. It has been said that software does not obey the laws of physics. Software will do whatever it has been programmed to do. Hence, unpredicted stimuli may have a more pronounced degree of disruptions than, say, hardware.

3.3.3 Hardware Agents

One would think that hardware agents would be relatively resistant to disruptions. Although hardware agents probably exhibit fewer disruptive actions than, say, humans or software, hardware agents still do fail. The vibrating struts in the Mars Polar Lander are an example of a hardware agent exhibiting disruptions behavior. The Tacoma Narrows bridge disaster described by Ketchum (n.d.) is an example of a purely hardware agent impacted by unpredicted aerodynamic forces. Hardware agents are strongly constrained by the laws of physics. So to that extent, hardware agents are predictable. However, the stimuli that act on the hardware are often difficult to predict. Therefore, the reaction of the hardware may be unpredictable also. Chapter 4, Case Histories, discusses both the Mars Polar Lander and Tacoma Narrows examples.

3.3.4 Number of Agents

Another way to describe disruptions is by the number of agents involved. That is to say, it is important whether an agent is acting alone or whether there is an interaction among agents.

3.3.4.1 Single-Agent Disruptions. A single-agent disruption results from a single component: human, software, or hardware. Figure 3.1 is a simple way to depict single-agent disruptions. First of all, assume that A is any agent within the system and that it is designed to perform a function F_A. The issue is the nature of the output (O) and whether this output is predictable. If it is not predictable, then the output can be said to be disruptive.

There are two ways that the output (O) can be disruptive. First, the input (I) can be unpredictable or random. As was said before, this is the case especially with human actions, such as pilot control of aircraft or of the efforts of humans in natural disasters or terrorist situations, such as the attacks on the towers on September 11, 2001, as described by Mendoça and Wallace (2006).

The second way that the output (O) can be a disruption is that the function F_A itself is unpredictable. This was the case in the Tacoma Narrows bridge collapse in which the aerodynamic aspects of F_A were unknown.

3.3.4.2 Multiple-Agent Disruptions. Disruptions often occur when two or more agents are in conflict. Each agent may perform the way it was designed; that is to say there was no individual component failure. The important thing is that it was the interaction of the agents that caused the disruptions. There is no limit to the number of agents involved in a multiple-agent disruption. When large numbers of agents are involved, a network failure occurs. Leveson (2008) refers to accidents that result from multiple agents as one type of "system" accidents.

Leveson (2002b, p. 149), for example, points out that normally operating components are usually ignored in traditional safety analysis. In Chapter 4, Case Histories, there are two accidents, the Mars Polar Lander and the Helios 522 accidents, in which system components operated exactly the way they were designed to operate. It was the interaction between these components that resulted in the accident.

Multiple-agent disruptions are depicted in Figure 3.2. The principal reason for multiple-agent disruptions is that the designers had no idea there was any interaction at all. Agents in this case do not have to be a single type of agent. For example, interacting agents could be a hardware–software combination, as

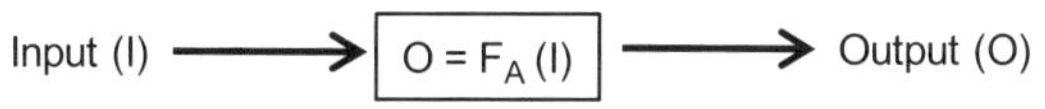

Figure 3.1. Single-agent disruptions.

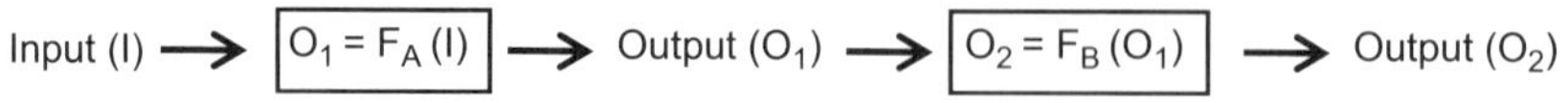

Figure 3.2. Multiple-agent disruptions.

in the case of the Mars Polar Lander, or a human-software combination, as in the case of Nagoya.

Consider, in this case, two agents performing two functions, F_A and F_B, as shown in Figure 3.2. In this case, the possibilities for disruptions are many. As in the case of single-agent disruptions, the input (I) might be predictable or random, in which case the flow-through to output (O_2) might be random and therefore disruptive.

However, the primary driver of disruptions in this case is the fact that the output (O_1) of F_A is also an input to F_B. If the designer of agent B is unaware of the fact that O_1 is an input to F_B, then O_2 may be highly disruptive.

This was the case in two notable accidents. First, in the Nagoya accident, the pilot (agent A) performed an inadvertent command; output O_1, on the flight control system (agent B). The results were catastrophic. In this case, agent A was a human agent and agent B was a software agent. This accident is discussed in Chapter 4, Case Histories.

The second example of multiple-agent disruptions is the Mars Polar Lander accident described by Leveson (2002). In this case, the vibration, O_1, was the output from agent A, the vehicle strut. Agent B was the vehicle software that interpreted the vibration as the vehicle having landed. Hence, output O_2 was a command to shut the propulsion down.

Another highly publicized example of two-agent weak disruptions was the Mars Climate Orbiter described by the National Aeronautics and Space Administration (NASA) Climate Orbiter website (1997). In this case, the fight control computer sent a signal to the thrusters in English rather than metric units. A rigorous interface management process would have avoided this problem.

Dekker and Hollnagel (2006) also point to the example of the Helios 522 accident as a multiple-agent disaster. In this accident, the crew and passengers died of hypoxic hypoxia before they could compensate for a pressurization malfunction. Hence, even though only one component failed, as a result of human error in settings, the pressurization system, the system was not able to compensate for the failure before the automatic flight control system took the aircraft to an unacceptably high altitude. Dekker and Hollnagel, therefore, refer to this accident as *systemic* rather than the fault of an individual component.

Gribble (2001) argues that the interaction among components is the primary source of fragility in modern complex systems. Gribble also argues that not much can be done about this interaction except to design for graceful degradation. Chapter 8, Resilience Architecting, shows that the *graceful degradation* heuristic is one of the primary ways to achieve resilience.

3.3.5 The Swiss Cheese Model

One of the more well-known models for depicting multiple sources of disruption is the Swiss cheese model. Reason (1997, pp. 11–13), for example, explains that the Swiss cheese model represents a series of defenses of any system. Figure 3.3 shows the Swiss cheese model with two layers of defense against disruptions. The first layer may be, for example, the normal technical measures that are built into a system to minimize the likelihood of disruption. For example, they may be redundant circuits or radiation shielding. The second layer may be the human backup needed for anticipation of unsafe conditions. Reason points out that each layer may have "holes." Each hole represents a threat to the safe operation of the system. The arrow represents the trajectory of a potential disruption. If the trajectory passes through all the holes, then a disruption will occur.

As discussed in Chapter 4, Case Histories, Chernobyl had two such layers. The nuclear power plant had the normal built-in safeguards. The "hole" was its sensitivity to rapid shutdown. The second layer was the humans who were responsible for watching the temperature in the reactor and taking the appropriate actions. The "hole" in this layer was the poor decision process that allowed the temperature to rise beyond the safe level. The combination of these two holes resulted in the well-known accident.

Reason notes that any system is vulnerable to these holes because the holes are dynamic. They shift in accordance with operating conditions, training of

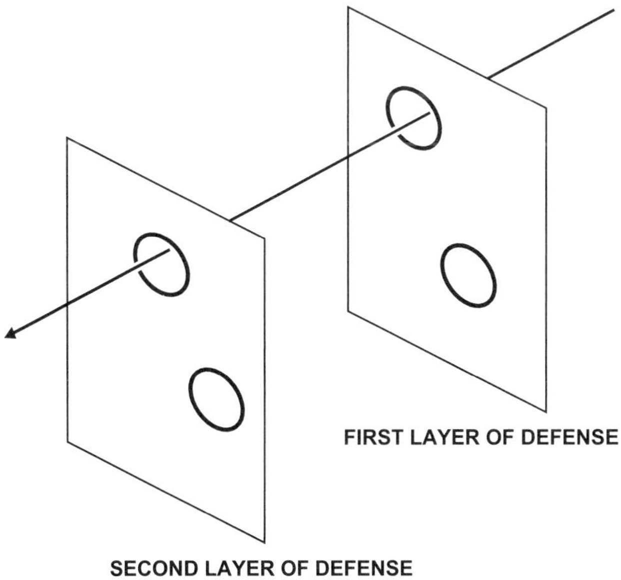

Figure 3.3. The Swiss cheese model. Adapted from Reason (1997, p 12).

operators, and other factors. The fact that these holes are moving makes the likelihood high that they will align and result in a disruption.

One conclusion to be drawn from the Swiss cheese model is that the systems with the most layers of defense are the most resilient. The converse of this conclusion is that systems with only a single layer of defense are the most brittle. For example, a system, such as a rail transportation system, which relies only on human operators to maintain safety, can be said to be brittle. Chapter 8, Resilience Architecting, discusses the *functional redundancy* heuristic. This heuristic says that there should be multiple ways to perform every critical function. Each one of these ways can be considered a layer in the Swiss cheese model. And the layers with the fewest holes contribute most to resilience.

3.3.6 Normal Accidents

Perrow (1999, p. 357) expresses an idea very similar to the Swiss cheese model. Perrow calls his theory the *normal accident* theory. He says that occasionally "two or more failures, none of them devastating in themselves, come together in unexpected ways and beat the safety devices." These "two or more failures" can be considered to be the holes in the Swiss cheese described above.

However, Perrow (pp. 89–94) goes on to add the coupling effect. Tight coupling, according to Perrow, happens when one failure has an immediate and direct effect on another system element. This type of coupling happens, for example, when a failure in a power grid causes a cascading failure throughout the grid system. Tight and loose coupling will be discussed in the depth below.

3.3.7 Tight and Loose Coupling

Disruptions can be either tightly or loosely coupled. Although Perrow does not use the terms *disruption* or *resilience*, he concludes that tight and loose coupling do have an influence on resilience, or as he puts it, the ability to recover. According to Perrow (pp. 89–90), tight coupling means that "there is no slack between or buffer between two given items." In engineering terms, there is tight coupling between almost all elements of a system. For example, if you put your foot on the accelerator, the car will gain speed. Perrow says that tight coupling can happen in organizational systems in which one process is strongly dependent on another process.

In loosely coupled systems, there is flexibility and tolerance. The airways system for example is loosely coupled. It can accommodate military, commercial, and private aircraft. Flights can be rescheduled, and landings can be changed. Therefore, loosely coupled systems are inherently more resilient. Perrow clarifies that all loosely coupled systems are not resilient, but, he says, it makes the job a lot easier. It can therefore be said that one purpose of architecting resilience, the subject of this book, is to loosen the coupling of the system elements.

3.3.8 The Multiple-Agent N^2 Diagram

Another way to view multiple-agent disruptions is through the multiple-agent N^2 diagram shown in Figure 3.4. The N^2 diagram is a well-known artifact in analytic analyses, but in the context of disruptions, the N^2 diagram takes on an added dimension of importance. The N^2 diagram is a simple method for showing how components interact with each other. In conventional methodology, these interactions are, for the most part, desirable, as designated by the letter D in the diagram. This chart shows, for example, that the interaction between Components 2 and 1 is desirable. It may be, for example, a simple electronic message.

However, the interaction may be undesirable, as designated by the letter U in the diagram. For example, the interaction between Components 3 and 1 is undesirable. It may, for example, cause excess heat or vibration. In Chapter 4, Case Histories, several examples of undesirable interactions are presented. In the case of the Mars Polar Lander and Helios 522 the undesirable interaction was the major disruption that eventually led to the catastrophes. In both cases, at least one component performed as it was designed to perform. That is, there was no component failure on the part of that component. In the case of the Nagoya accident, the same was true. The pilot made an inadvertent but innocent mistake, and the flight control system performed as designed. These cases are discussed more thoroughly in Chapter 4. However, the lesson to be learned from these cases is that it was not the failure of the individual components that resulted in disruption, but rather the *interaction* between the components. Following are some additional observations on the validity and use of the multiple-agent N^2 diagram.

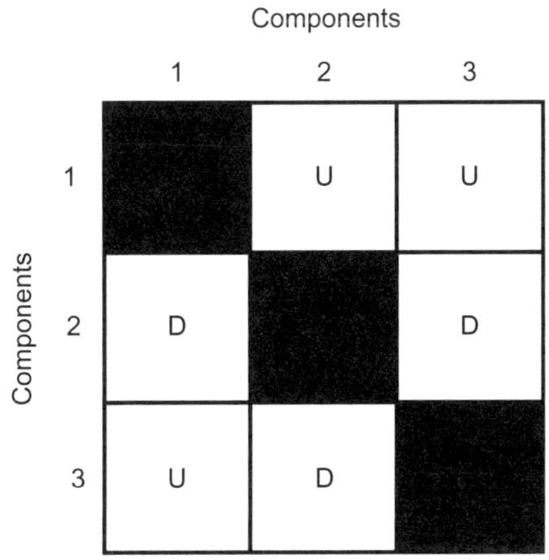

Figure 3.4. The multiple-agent N^2 diagram.

Perrow (1999, p. 98) refers to these undesirable interactions as "negative synergy." Perrow also goes on to define another phenomenon he calls the "Union Carbide factor" (p. 356) after the company that owned the Bhopal chemical plant. The essence of the Union Carbide factor is that the interactions have to be both undesirable and just the right interactions to cause a catastrophe. In other words, there may be many undesirable interactions, but only certain interactions will result in catastrophic accidents.

First, as a classic analytic methodology, the N^2 concept is somewhat simplistic. It assumes, for example, that all desirable and undesirable interactions occur between individual pairs of components. In a truly complex system, the interaction may require three or more components to create a meaningful interaction. Mathematical methods do exist for depicting N-dimensional interactions, tensor analysis, for example. Implementing such methods, although mathematically sound, may be difficult. It is suggested that if the one-on-one interactions are thoroughly analyzed, progress will have been made in identifying interactions.

As an alternative to creating a large N^2 diagram for an entire system, it might be more practical to create sub-N^2 diagrams. This approach would be most useful on aircraft or spacecraft systems. A meaningful sub-N^2 diagram might be, for example, a matrix of all the components on the systems that are electromagnetic interference (EMI) emitters, such as solenoids, with all the components that are vulnerable to EMI.

Finally, the N^2 diagram is only a method of identifying *predictable* interactions. It will do little to predict interactions that may develop, for example, from complex software algorithms. Predicting disruptions that originate from multiple-agent interactions is only one aspect of resilience, not the whole story. Although predicting and dealing with interactions is one way to prevent disruptions—and this process is value added—it is also necessary to look at all the other ways to make a system resilient as discussed in Chapter 8, Resilience Architecting.

3.3.9 Disruptions and the Law of Large Numbers

As a final note on multiple-agent disruptions, it is useful to reflect on the law of large numbers. This law, well known in statistics, simply says that as the number of trials of a random variable increases, the mean value of the result will approach a fixed or stable value.

This law is particularly important when the number of possible multiple-agent disruptions is considered. Although the probability of a particular event may be very small, and the probability of a combination of events may be infinitesimally small, the probability of *any* combination of events may be significant. Thus, the number of trials increases as the number of combinations becomes larger. And hence, the probability of an event, such as a disruption, will approach a number determined by the probability of a single event and the number of combinations.

Utts and Heckard (2005), for example, cite the example of a person winning the lottery twice, which, they point out, has happened. It only seems unlikely if you ask what the probability is of a particular person winning the lottery twice. However, if one asks what the probability is of any person, anywhere, winning the lottery twice, the answer is much greater. As a matter of fact, Utts and Heckard say that there are "better than even odds that there would be a double winner in a seven-year period somewhere in the United States."

So what does winning the lottery twice have to do with disruptions? The answer is quite a lot. Say, for example, that any two functions in a system, when performed as designed, are benign; that is, they do not create a disruption. However, when they interact, their interaction may constitute a disruption. If the number of such non-benign interactions is large, then the likelihood is that at least one of these combinations of events will result in a disruption, just as in the example of winning the lottery twice.

3.4 PREDICTED VERSUS UNPREDICTED DISRUPTIONS

Traditional system safety analysis, as documented in *MIL-STD-882* (1993), depends for the most part on designing systems to survive disruptions that have happened before. The premise is that a system designed for resilience will avoid, survive, and recover from disruptions that are completely unpredicted.

Many disturbances are predictable, at least on a probabilistic basis. Hurricanes on the Gulf Coast, earthquakes in California, and tornadoes in Kansas happen on a regular basis, so their frequency of occurrence and magnitude are well known.

Unpredicted disturbances of all three types can occur, either because a phenomenon was unknown to science in general or because it was unknown to the designers of a system.

The disruptions caused by the turbulent boundary layer effect in the Tacoma Narrows bridge collapse of 1940 is an example of a disruption caused by a phenomenon unknown to the designers. However, the Van Allen Radiation Belt, whose existence was not even known, is an example of a phenomenon unknown to the entire scientific community. This phenomenon is described by the Encyclopedia Britannica (2008).

The source of unpredicted disruptions is often, but not always, human. As mentioned before, despite extensive training, humans can often perform completely unpredictable acts. These acts are not necessarily a lack of competence or a desire to do evil. They can be completely innocent mistakes. Woods (2007) says that "it is fundamental in the universe that you can't anticipate everything."

Strong disruptions do not depend on the quality of the engineering that is employed. Because its causes are completely random, it will occur whether the quality of the engineering is high or low. The quality of engineering will come into play in determining whether the system can survive the strongly disruptive event.

An obvious question is as follows: If disruptions are unpredictable, then why study them? The question is not how to predict them, or even to eradicate them. The question is how should they be dealt with? The answer lies in the holistic approaches that were discussed in Chapter 6, Capabilities.

3.5 IMPLICATIONS FOR RESILIENCE TO DISRUPTIONS

The methods most appropriate for increasing the resilience of a system will result from the types of disruptions that are expected. These methods fall into the categories of analytic and holistic methods to be described more fully in Chapter 6, Capabilities. These methods are described briefly in the discussion of the systems approach in Chapter 2, System Resilience and Related Concepts.

3.5.1 Analytic Methods

Traditional analytic methods can be brought to bear in many disruptive situations, particularly with respect to predicted disruptions. The difference is that traditional processes, such as requirements, verification, and interfaces, would have to be treated at a level of detail not common in industry today. Hence, the first aspect of analytic methods is that they would have to be enhanced in *increased detail*. However, the need for this increased detail is determined by factors of possible consequences, risk, cost, and benefit-cost ratio as discussed in Chapter 12, Implementation.

With respect to the Mars Polar Lander, for example, the requirements for strut vibration and software logic could have indeed been developed to a level of detail to reduce the likelihood of the failure that occurred as a result of the weak disruptions. In addition, simulation of the strut and software could have been developed to verify those requirements by analysis.

Another analytic method important in the Mars Polar Lander case is interface management. Although interface management normally pertains to planned, or desirable, interfaces, weak disruptions' causes are unplanned, or undesirable, interfaces. This effect is particularly important in the two-agent scenario. If all possible interfaces were examined for the possibility of weak disruptions, many of these situations could be predicted.

This example is completely compatible with the Woods (2007) heuristic, "avoid hidden interactions." In more traditional phraseology, this could be reworded as "avoid undesirable interfaces."

In addition to detail, another characteristic of analytic methods that has the potential for dealing with weak disruptions is *breadth*. Analytic methods, as practiced, are often limited to one product system, for example, a commercial aircraft, and one organization, for example, the developer. Developers do attempt to extend these principles to their suppliers who make components and subsystems. However, extension of analytic principles to maintenance and production systems is rare, especially when the maintenance organization, for example, is separated from the developer by multiple organizational boundaries. Such was the case in the ValuJet accident, described by Chiles (2002, p. 310), in which the maintenance organization was separated from the developer by two layers or organizational boundaries. Rigorous extension of requirements and verification across these boundaries is essential for resilience.

3.5.2 Holistic Methods

The word *holistic* is used to describe methods that are comprehensive in nature as opposed to the analytic methods described above that are based on flowing down requirements to individual components of a system. Holistic methods can be considered to cover such tools as heuristics, risk analysis, and cultural initiatives.

Heuristics can be applied to almost every type of disruption. We saw that the heuristic of tolerance was particularly important for Type A disruptions of input. Billings' heuristics of human-machine interfaces are important for Type B disruptions of loss of function, capability, or capacity. Typical of these heuristics is "the human must always be in control." We saw that for Type B disruptions of environment, the heuristics of margin and cross-scale interactions were particularly important.

3.6 FURTHER EXPLORATION

1. Look at some of the case histories in Chapter 4, Case Histories. Decide whether the disruptions were Type A, external, or Type B, internal. Explain the rationale for your choice.
2. Look at the case studies of Chapter 4, Case Histories, and identify which ones fall into the category of multiple-agent disruption. Explain your answer. What were the agents, and what was the nature of their interaction that caused the disruption.
3. Examine the existing literature on traditional systems engineering and suggest ways that its scope can be expanded to avoid disruptions.
4. From your research, identify examples of single-agent and multiple-agent disruptions. Discuss.

Chapter **4**

Case Histories

It is not the purpose of this chapter simply to repeat the case histories and root causes of the major accidents documented in other sources. Rather, it is the purpose to show the commonality among these root causes to identify the attributes and capabilities required to create a system resilient to these types of accidents. Hence, this chapter summarizes the root causes using the categories used throughout this book, in particular those that stimulate a necessary capability in Chapter 6, Capabilities. Although the cases listed here may not be a comprehensive list, it is apparent that they have root causes in common regardless of the domain, for example, an infrastructure or a chemical power plant disaster. It is this commonality among root causes that allows the analyst to arrive at common approaches to accident avoidance, survival, or recovery. In addition, expressing these root causes in a language common to system resilience aids in showing this commonality.

Some analysts list the root cause of some disasters as design errors. This chapter takes these cases down a layer to more fundamental causes. Design errors can, for example, result from a failure to identify requirements, to verify that the system met those requirements or that the designers did not adequately consider the technical risks. Reason (1997, pp. 10–11) refers to flaws at this stage of a system's life cycle as "latent errors." For example, Chiles (2002) would agree that there were no major design flaws in the Chernobyl reactor. However, the designers failed to recognize that the design was highly sensitive to control variations. Therefore, a technical risk existed that the designers did not properly address. Hence, the errors that matter are systemic errors, as opposed to symptoms, as discussed by Leveson (2004a).

Other root causes could simply be classified as cultural factors. Vaughn (1996), for example, describes how the National Aeronautics and Space

Administration (NASA) culture was one within which deviance from desired procedures and desired performance was recognized to be the norm. Cultural factors can also be considered to be systemic factors because they affect all elements of a system. Although all of these researchers are in basic agreement about the causes of these disasters, you will observe that their points of emphasis often diverge.

The material in this chapter is taken completely from the public literature. Furthermore, there is general agreement among the sources on the causes of the various accidents. Finally, the sources bring a degree of expertise and analysis to the study of these accidents that were not addressed by the investigation teams. This analysis is particularly valuable to our study of resilience.

Finally, this chapter goes beyond the official reports on these accidents to interpret them in the light of resilience attributes, such as capacity, flexibility, tolerance, and inter-element collaboration, as defined in this book. The original sources do not, in general, use these terms. It is hoped that this interpretation will shed greater light on these cases.

Chapter 8, Resilience Architecting, elaborates on these attributes, but basically they are as follows: Capacity asks whether the system has enough margin and resources to absorb a disruption. Flexibility asks whether the system can perform the function in an alternative manner. Tolerance asks whether the system will degrade slowly when subjected to a disruption. Inter-element collaboration asks whether there is communication and collaboration among the system elements.

4.1 THE CASE FOR CASE HISTORIES

One might ask why we should have a chapter on case histories at all because many thinkers and researchers within the resilience community itself believe that catastrophes are so random that they cannot be forecasted. Indeed, as we saw in Chapter 3, Disruptions, there is a whole category of disruptions called unpredicted disruptions. Many observers, namely Leveson et al. (1995, p. 76) and Reason (1997, pp. 118–119) have observed that almost all catastrophes are preceded by "near misses" and "incidents," which may have been an indicator accident potential. Leveson and Reason also suggest that the time is ripe for research into the value of near misses and incidents for assessing accident potential.

Wright and Van der Schaaf (2004) have made significant gains in proving this hypothesis. As discussed in Chapter 10, Measuring Resilience, they have shown that there is a strong correlation between the causes of near misses and major accidents. This correlation is known as the "iceberg theory."

There is a second reason for studying case histories and root causes: This study gives us an insight into how a disruption can be survived even if it is not avoided. The two most notable examples are Apollo 13 and the New York Power Restoration cases, which are both discussed in this chapter and analyzed in Chapter 8, Resilience Architecting.

A third and compelling reason for studying case histories is that they provide the scenarios for system definition. Chapter 8 discusses the need for defining the scenarios in which we define the system. The heuristic employed is that the analyst should define the system using, first, the worst case scenario, and secondly, the most likely scenario. Chapter 8 discusses the logic for using this heuristic.

Finally, case studies provide the basis for risk analysis, for the basis of risk analysis is that observing past events will provide the basis for avoiding future potential accidents.

4.2 CASE STUDIES

In the representative case histories to follow, a table is provided for each one that provides some of the resilience strengths or weaknesses of each system being studied. These strengths and weaknesses are expressed in terms of either the resilience attributes listed in Chapter 2, System Resilience and Related Concepts, or in terms of specific analytic processes that may have been neglected, such as requirements, verification, reliability, or interface management. No attempt is made here to redesign the failed systems. However, Chapter 8, Resilience Architecting, does provide a set of heuristics that can be used to design more resilient systems in the future.

Associated with each case study is a small table that summarizes some of the resilience aspects for each incident or accident. It should not be inferred that these aspects constitute all the aspects that would have needed to be addressed to make the system resilient. They only constitute the conclusions of the investigation for each incident. Chapter 8, Resilience Architecting, will summarize the heuristics, or design practices, that would make each system more resilient. Whether these practices could have been implemented in the cases discussed in this chapter is dependent on constraints associated with those cases.

Among the case studies below most are of catastrophic disasters. However, among these are a few incidents in which the system was found to demonstrate resilience and therefore either survived completely or recovered with a certain degree of performance. Apollo 13, the New York Power Restoration, and U.S. Airways Flight 1549 are examples. The reason these are shown is that they can shed as much light on resilience as the disasters can.

It will also be noted that most of the cases cited in this chapter are the result of Type B, that is, systemic, disruptions rather than Type A, that is, disruptions of input. This is not to say that most disasters result from Type B disruptions. This may be the case, but that conclusion was not intended here.

Finally, at the time of this writing, the official reports had not been written for certain accidents, in particular Metrolink 111 and U.S. Airways 1549. However, despite the lack of official reports, it is still possible to draw conclusions from accounts in the public media about the resilience of the systems involved.

4.2.1 Challenger

On January 28, 1986, the space shuttle Challenger failed 73 seconds into flight as a result of hot gas penetrating a seal causing a strut to fail. Most sources, primarily Vaughn (1996), focus on the pressures to launch and the capitulation by the propulsion contractor, Morton Thiokol, to these pressures. However, the causes are more complex. No accident has provoked more debate and speculation about its root causes than Challenger. This accident is also discussed by Leveson (1995, pp. 569–580) and Chiles (2002, pp. 65–68, 72–79, 84–93, 307).

The failure of the O-rings was identified as the immediate cause of the disruption making this a Type B, or systemic disruption. However, this was not a failure of unreliability; it can better be classified as a failure of verification because the O-rings had never been tested at the temperatures they experienced that day. The failure to qualify the O-rings at that temperature makes the failure one of management rather than technical. The failure can also be classified as multiagent, as defined in Chapter 3, Disruptions, because nearby components failed as a result of the hot gases that escaped through the O-rings (Table 4.1).

Reason (1997, pp. 157–160), for example, focuses on the failure of the regulatory process to maintain its independence. In essence, NASA regulated itself by a set of flawed standards in which, according to Vaughn (1996) deviance became the norm, that is to say there was the philosophy of acceptable risk. Vaughn introduced a phrase intended to capture the mindset at NASA that, in their judgment, contributed to the disaster. The phrase was "normalization of deviance." This phrase implied a lack of concern for small problems that could eventually have catastrophic results. According to Vaughn, this lack of concern did indeed exist.

The main emphasis of Leveson (1995, pp. 569–570) was the accelerated flight schedule, budget cutbacks, and reductions in staffing, especially in safety. All of these factors added up to increased risk for the Challenger mission.

Table 4.1. Some Resilience Aspects of Challenger

Disruption: Type B—Systemic	
Inter-element collaboration	Deficient
Decision Making	Deficient
Cultural Factors	Deficient
Risk Management	Deficient
Verification	Deficient
Regulatory Process	Deficient
Management Oversight	Deficient
Capacity	Deficient
Flexibility	Deficient
Tolerance	Deficient

In addition, Paté-Cornell and Fischbeck (1994) lists three root causes: First, there was the miscommunication of technical uncertainties, especially between Morton Thiokol and NASA. Second, there was the failure to use information from past near misses, that is, from past missions in which the shuttle was allowed to fly without full readiness. Finally, there was the error in judgment in placing schedule before safety. Vaughn says that one engineer was asked to "take off his engineering hat and to put on his management hat."

Vaughn (1996, pp. 409–410) concludes that "...no rules were violated; there was no intent to do harm. Yet harm was done. Astronauts died."

According to Vaughn (1996, pp. 7–11), the Presidential Commission summarized the causes as follows: First, there was an inadequate safety process. The process failed to assign criticality categories to key components (the O-rings). The system safety process also failed to identify areas where redundancy was assumed to exist and it did not, as pointed out by Vaughn (1996, pp. 133–134). Leveson et al. call the O-ring problem an inattention to system control because the function of the O-rings was to control the release of the propellant gases. The safety process also failed to compile and disseminate trend data.

Next, there was a lack of a problem-reporting system. This is a common theme. Central to this issue was the fact that there were no members of NASA's safety body at the teleconference that resulted in the final decision.

Vaughn lists the additional root causes: There was the philosophy of "acceptable risk." Rather than a flaw in the risk-management process, this factor falls in the cultural paradigm category. In Chapter 5, Culture, this is called the *risk denial* paradigm. Leveson et al. (2006, p. 97) define risk in a broader sense, that is, as the "control" factor that maintains resilience within the organization. According to Leveson, this control factor was not present on Challenger.

Morton Thiokol engineers were not present at the launch. This factor violates the inter-element collaboration required by resilience, discussed in Chapter 8, Resilience Architecting.

Next, there was a lack of executive involvement in safety issues. This phenomenon is called the "distancing effect" as discussed in Chapter 5, Culture.

There was a shortfall in basic engineering processes, such as the lack of knowledge of launch requirements and incomplete testing.

4.2.2 Texas City—1947

On April 16, 1947 the French ship *Grandcamp* carrying ammonium nitrate (the same explosive as in Oklahoma City) exploded in the port of Texas City, Texas killing almost 600 people, the largest industrial accident in U.S. history. Stephens (1997), Chiles (2002, pp. 216–217, 299), and Perrow (1984, pp. 105–108) cite three organizational problems, two of them regulatory in nature. First, the U.S. Coast Guard, which has the official responsibility for controlling dangerous material entering U.S. ports, operated in a passive mode, that is, they made no attempts to determine whether any ships carried dangerous cargo unless they were alerted. Hence, the Coast Guard was completely ignorant of the cargo.

Second, the port authority of Texas City, which had the responsibility for controlling smoking on the docks, was lax in enforcement.

So the combination of the lack of coordination among the ship staff, port authority, and the Coast Guard is the key aspect to the deficiency in interelement collaboration required by resilience.

Finally, the French captain ordered the hatches of the ship closed in the misguided belief that it would snuff out the fire that preceded the explosion, which was another error in judgment.

Perrow (1999, pp. 106–108) also mentions another explosion in a Texas City petrochemical plant in 1969. In the latter case, the crisis was brought under control by swift action of the operators, and no fatalities occurred.

Some resilience deficiencies are summarized in Table 4.2.

4.2.3 Texas City—2005

On March 23, 2005, the BP Texas City refinery experienced a process accident that resulted in 15 fatalities and more than 170 injuries. The report of the independent panel headed by Baker et al. (2007) concluded that there were deficiencies in safety management, safety oversight, and safety culture. Of these three, culture received the major emphasis in the report. Chapter 5, Culture, discusses the approach to culture recommended by the panel for BP. Some resilience deficiencies of the Texas City 2005 accident are summarized in Table 4.3.

In addition, the lack of coordination between management and safety deprived the refinery of the interelement collaboration required by resilience.

4.2.4 Piper Alpha

The North Sea oil and gas platform, known as Piper Alpha, exploded in 1988 killing 165 of the 226 people on board. Reason (1997, pp. 161–163) lays the lion's share of the blame on a faulty maintenance system. This is just one of several catastrophes for which faulty maintenance was the major cause. The

Table 4.2. Some Resilience Aspects of the Texas City—1947 Disaster

Interelement collaboration	Deficient
Decision Making	Deficient
Regulation	Deficient
Capacity	Deficient
Tolerance	Deficient
Flexibility	Deficient

Table 4.3. Some Resilience Aspects of the Texas City—2005 Accident

Disruption: Type B—Systemic	
Culture	Deficient
Managerial Oversight	Deficient
Capacity	Deficient
Flexibility	Deficient
Tolerance	Deficient
Interelementcollaboration	Deficient

others include Apollo 13, Bhopal, and Three Mile Island. This accident is also discussed by Chiles (2002, pp. 245–247, 308) and Paté-Cornell (1990).

On the Piper Alpha, a night maintenance crew had attempted to restart a pump that had been shut down by the day shift and a safety valve removed. The result was a leak in condensate that exploded.

Causes like this can be put in a larger category of communication failures like that of Challenger. In summary, the causes are as follows.

First, there was a lack of communication at the detail (maintenance) level. This lack of communication was the primary contributor to the lack of interelement collaboration required by resilience, as discussed in Chapter 8, Resilience Architecting.

Next, there was a complete lack of regulatory oversight. This shortfall illustrates a basic required capability as described in *Regulatory Environment* in Chapter 2, System Resilience and Related Concepts.

Next, there was lack of attention to safety in operations. For example, there were no lifeboat drills, inadequate fire containment, corroded sprinklers, wrong procedures (people who died mustered in correct area: those who survived did the wrong thing), or cursory inspections.

Finally, there were conflicting priorities. Flin (2006, p. 226) cites Piper Alpha as an example of a conflict between pressures for improved production and the objective to reduce costs. Flin also points to another conflict in priorities when adjacent platforms continued to pump oil to Piper Alpha even after the fire had started. Chapter 5, Culture, shows conflicting priorities to be a basic culture deficiency.

Table 4.4 summarizes some of the resilience aspects of Piper Alpha.

4.2.5 Columbia

The Columbia Accident Investigation Board report (2003, p. 11) states that the Columbia space vehicle was a compromise design that "never met any of its original requirements for reliability, cost, ease of turnaround, maintainability or, regrettably, safety." See Chapter 6, Capabilities, for discussions of reliability, maintenance and safety.

Table 4.4. Some Resilience Aspects of the Piper Alphas Disaster

Disruption: Type B—Systemic	
Inter-element collaboration	Deficient
Decision Making	Deficient
Cultural Factors	Deficient
Capacity	Deficient
Flexibility	Deficient
Tolerance	Deficient
Safety	Deficient
Regulatory oversight	Deficient
Culture	Deficient

The final flight of Columbia on February 1, 2003 resulted in the loss of the space vehicle and its seven-member crew. During lift off, there were no indications to engineers or to Mission Control that the foam strike during ascent had caused any problems at all. There were some weak signals, but they went undetected by Mission Control. The Columbia Accident Investigation Board (CAIB) concluded that the foam had caused superheated air to enter the wing's leading edge support structure eventually causing the Orbiter (the reentry vehicle) to fall out of orbit and catastrophically fail.

A key finding of the report was that culture (See Chapter 5, Culture) had as much to do with the accident as the foam. The Board found that NASA's organization (1) does not have effective checks and balances (See Chapter 9, Governance), (2) does not have an independent safety program (See *Safety* in Chapter 6, Capabilities), and (3) has not demonstrated that it is a learning organization (See *Commitment to Resilience and a Learning Culture* in Chapter 5, Culture). Chapter 11 of the report makes no specific recommendations for cultural change. Most changes relate to technical and managerial processes. Chapter 5 of this book provides some suggested approaches for cultural change.

4.2.5.1 Recommendations. Following is a summary of the recommendations couched in terms of the terminology used in this book as reflected in the Columbia Accident Investigation Board report (2003).

Most of the technical recommendations fall into the following categories: First, there is technical risk. The report makes many recommendations designed to reduce technical risk. Most of these had to do with the ability of the Orbiter to withstand minor debris damage or failure to achieve orbit.

The report makes several recommendations pertaining to verification by inspection of the Orbiter at various phases of the mission. The report also recommends verification by analysis using physics-based models. The report also recommends increased testing.

The report recommends that schedules be set that are consistent with available resources. Once again, this recommendation is consistent with the attributes described in the Section 6.3.6 Schedule Management section of Chapter 6, Capabilities.

The report recommends training in various areas. This topic comes under the heading of Section 6.4.2.7 Expertise in Chapter 6, Capabilities. One significant recommendation is that the training should be across NASA and contractor lines. This recommendation is in agreement with the theme of this book that system resilience should be implemented across the entire infrastructure of a program. See Chapter 7, Infrastructure. Training is an aspect of Expertise covered in Chapter 6, Capabilities.

The primary organizational recommendation is an increased degree of independent review. The second recommendation is an increased degree of authority by the safety organization. Both of these recommendations are in line with the attributes expounded in Chapter 9, Governance. The final organizational recommendation is the integration of the entire program. This recommendation is completely in line with the concept of an integrated infrastructure discussed in Chapter 7, Infrastructure.

The report provides no recommendations pertaining to the cultural deficiencies it noted.

4.2.5.2 Root Causes. Although the Columbia report does not have a root cause summary, the following can be inferred from the above.

There was a lack of an attention to risk, technical, both technical and managerial, and especially to risks caused by schedule pressure. See the Risk Management section of Chapter 6, Capabilities. In particular the *risk denial* paradigm, as described in Chapter 5, Culture, was wide-spread.

There was a lack of verification, including testing, inspections, and physical modeling as discussed in Chapter 6, Capabilities.

There was a lack of expertise and trained people in all nodes of the program organization, including contractors. See the Section 6.4.2.7 Expertise section of Chapter 6, Capabilities.

There was a lack of program integration and a central authority responsible for the whole program. See Chapter 7, Infrastructure. This lack of program integration resulted in a deficient degree of interelement collaboration required by resilience as discussed in Chapter 8, Resilience Architecting.

There was a lack of independent review and authority for safety-related decisions, as discussed in Chapter 9, Governance.

One of the aspects of Columbia relative to resilience was the lack of external imaging during the mission. External imaging was offered but was declined. The Columbia Accident Investigation Board report (2003, p. 226) provides four separate recommendations for external imaging. This recommendation is relevant to the *drift correction* heuristic discussed in Chapter 8, Resilience Architecting. This heuristic calls for as much external information as possible to allow a system to anticipate a disruption and to avoid it if possible.

So, in general, from a resilience point of view, Table 4.5 provides a summary of some of the resilience aspects of Columbia.

Table 4.5. Some Resilience Aspects of Columbia

Disruption: Type B—Systemic	
Cultural factors	Deficient
Design flexibility	Deficient
Design tolerance	Deficient
Risk management	Deficient
Verification	Deficient
Management oversight	Deficient
Expertise	Deficient
Defect detection and correction	Deficient
Capacity	Deficient
Interelement collaboration	Deficient

4.2.6 Chernobyl

In April 1987 the V. I. Lenin nuclear reactor at Chernobyl in the then Soviet Republic of the Ukraine exploded with considerable release of radiation and loss of life. According to both Reason (1997, pp. 15–16, 76–77) and Chiles (2002, p. 307), Chernobyl was a disaster almost totally attributable to decision error. In fact, technically, nothing was really wrong with the reactor except for its intolerance to human error. The first decision error was the operators' choice to continue testing even though the power had fallen to a low level. The second error was the decision to cut off safety systems to continue the experiment. Leveson (1995, pp. 640–647) also points to weaknesses in the design of the reactor that make it harder to control.

4.2.6.1 Root Causes. The root causes of the Chernobyl disaster boil down to several factors.

There was a serious lack of decision-making expertise on site. See the discussion of decision making in Chapter 6, Capabilities. The management system that contributed to the deficient decision-making process is responsible for the lack of interelement collaboration required by resilience, as discussed in Chapter 8, Resilience Architecting.

There was a lack of design for technical risks associated with the reactor control. See Section 6.3.4 Risk Management in Chapter 6, Capabilities. The reactor design also failed the *gradual degradation* design requirement as discussed in Chapter 8, Resilience Architecting.

These risks are summarized in Table 4.6.

In addition, Chernobyl can be classified as a Type B, or systemic, disruption as defined in Chapter 3, Disruptions. In short, the failure of Chernobyl was an unexpected characteristic of the reactor caused by the unpredictable actions of humans.

Table 4.6. Some Resilience Aspects of Chernobyl

Disruption: Type B—Systemic	
Decision Making	Deficient
Design Tolerance	Deficient
Risk Management	Deficient
Capacity	Deficient
Flexibility	Deficient
Inter-element collaboration	Deficient

4.2.7 Bhopal

On December 3, 1984 a chemical factory in Bhopal, India released a cloud of methyl iso-cyanate (MIC). According to Chiles (2002, pp. 260–268, 306), the estimated fatalities are 7000. Although the entire chain of events is in dispute, there are some areas of agreement. What is not in dispute is that water got into the MIC tank. How it got into the tank is in dispute. Was it accidental? Was it sabotage? Was it put there intentionally to cool it off? No one will ever know. Assuming the failure was basically maintenance related, Bhopal can be considered the result of a Type B, or systemic disruption. However, if sabotage was the cause, then it would be a Type A, or external disruption. However, most authorities, Reason (1997, p. 54), Chiles (2002), Leveson (1995, pp. 598–608), and Perrow (1984, pp. 355–356) all agree that bad maintenance, in the form of a failure regularly to clean the pipes and valves, was the basic cause of the disruption.

Whatever the cause of the water influx, there was a basic design flaw that can only be described as an inadequate approach to risk mitigation. The flaw, according to Reason (1997, p. 54) was that there was an assumption of independence among the various subsystems involved in the disaster. Reason cites a lack of "defense in depth" as a cause. This cause evokes Reason's Swiss cheese model discussed in Chapter 3, Disruptions. This model shows that the more layers of defense a system has, the safer it will be.

Both Reason (1997) and Chiles (2002) agree that flawed decisions also contributed to the accident, namely the decision to disable safety systems. In summary, based on the findings of various researchers, the root causes include inadequate risk mitigation, inadequate safety safeguards, and poor maintenance.

Leveson (1995, pp. 599–600) describes the deterioration in expertise during the transition from U.S. to Indian control. Initially, the plant was operated by U.S. personnel. Then, there was a transition from U.S. personnel to Indian personnel who had been trained by the Americans. Eventually, the Indian personnel were replaced by more Indian personnel who were less well trained. At each step of the process, the expertise suffered.

Perrow (1999, pp. 355–356) classifies Bhopal as an example of a "system failure," as opposed to a "component failure" for the reason that just the right combination of events had to happen for disaster to occur. Perrow (p. 356) calls this right combination of factors the "Union Carbide" factor.

Hence, the lack of cooperation among management, safety, and maintenance was the core issue responsible for the lack of interelement collaboration required by resilience. Table 4.7 summarizes some of the resilience aspects of Bhopal.

4.2.8 Three Mile Island

On March 28, 1979, the Three Mile Island nuclear reactor near Middletown, Pennsylvania, experienced partial meltdown of the reactor core when reactor coolant escaped for 2 hours and 20 minutes after a relief valve stuck in the open position and operators cut back on emergency cooling.

According to Reason (1997, pp. 87–88), the root cause of the Three Mile Island accident was not one, but two different maintenance errors. The first maintenance error was to leave valves closed so that heat could not be removed. The second maintenance error was when another valve stuck in the open position causing radioactive water to flow into the containment area and then into the basement. According to Perrow (1999, p. 19), much of the Presidential Commission's attention was focused on finding out who was responsible for leaving the relief valves closed. The answer to this question was never determined.

With regard to the maintenance errors, Leveson (1995, pp. 623–629) points a finger at the failure of the operators to heed clear warning signs. However, the Kennedy Commission noted that the operators were poorly trained and therefore not likely to recognize the warning signs or to know what to do when they happened. These warning signs and their correction call for the employment of the *drift correction* heuristic to be discussed in Chapter 8, Resilience Architecting.

Chiles (2002, pp. 45–50, 304) points to two different decisions, one of which made matters worse and one kept the situation from getting worse than it was.

Table 4.7. Some Resilience Aspects of Bhopal

Disruption: Type B—Systemic	
Risk management	Deficient
Maintenance	Deficient
Design tolerance	Deficient
Decision making	Deficient
Management oversight	Deficient
Capacity	Deficient
Flexibility	Deficient
Interelement collaboration	Deficient
Expertise	Deficient

The first is an example of Cognitive Lock in which a maintenance worker continued to perform incorrect operations that contradicted safety policy and evidence from the scene. The other worker was able to cut off critical systems to reduce the damage. This accident is also discussed by Leveson (1995, pp. 619–639) and Perrow (1984, pp. 15–31).

None of the sources points to serious design errors except for the lack of visibility of key instruments. This lack of visibility and the lack of communications among maintenance workers all constituted a deficiency in interelement collaboration which is a key attribute of resilience.

Apart from all the above, Perrow (p. 16) points to negative cultural factors. Three Mile Island, he says, points to a "seemingly endless story of incompetence, dishonesty, and cover-ups before, during, and after the event."

In summary, the primary root cause was a faulty maintenance system. See the Section 6.4.2.13 Maintenance in Chapter 6, Capabilities. Table 4.8 summarizes some of the resilience aspects of Three Mile Island.

4.2.9 Clapham Junction

On December 12, 1988, multiple trains collided near Clapham Junction station in England killing 35 people and injuring 500. The investigation concluded that a technician had incorrectly wired a signal box.

According to Reason (1997, p. 90), this is a classic case in which a highly motivated person makes a mistake often dismissed as "human error." However, the inquiry concluded that the problem was systemic rather than individual because the technician had not received the proper training for this job.

So the root cause can be classified as lack of expertise, which is discussed further in the Section 6.4.2.7 *Expertise* of Chapter 6, Capabilities. Reason (1997) also notes the lack of supervisory checks that contributed to this accident. This lack of supervisory checks contributed to the deficiency in interelement collaboration, which is an attribute of resilience.

Table 4.8. Some Resilience Aspects of Three Mile Island

Disruption: Type B—Systemic	
Maintenance	Deficient
Defect Visibility	Deficient
Decision Making	Mixed
Capacity	Adequate
Flexibility	Adequate
Tolerance	Adequate
Inter-element collaboration	Deficient
Culture	Deficient

Table 4.9. Some Resilience Aspects of Clapham Junction

Disruption: Type B—Systemic	
Expertise	Deficient
Management oversight	Deficient
Decision Making	Deficient
Capacity	Not applicable
Flexibility	Deficient
Tolerance	Deficient
Interelement collaboration	Deficient

In addition, Clapham Junction could be classified as a Type B, or systemic, disruption, as defined in Chapter 3, Disruptions, because of the unpredictable action of humans who were components of the system. Table 4.9 summarizes some of the resilience aspects of Clapham Junction.

4.2.10 TWA 800

On July 17, 1996, Trans World Airlines (TWA) flight 800, a Boeing 747-131, crashed near New York killing all 230 people on board. The immediate cause was a fire in the center wing fuel tank (CWT). Although there is some uncertainty in the source of the ignition, the National Transportation Safety Board (NTSB) (2000) concluded that the most likely source was a short circuit in the wiring associated with the fuel quantity gauging system.

Although there were reports of a missile attack that have not been verified, the NTSB solely concerned itself with the fuel tank flammability issue.

From a resilience point of view, the NTSB concluded that a contributory factor was the design and certification concept that assumed that all ignition sources could be identified and precluded by design methodologies. Furthermore, the design and certification concept had no methodology to render the fuel vapor nonflammable. In resilience terminology, it can be said that there was inadequate capacity and tolerance.

Hence, the ultimate root cause was the failure to identify the risks associated with fuel vapor and to incorporate the results into the design. Beyond the failure to have a risk process, the real failure is to have methodologies that could realistically estimate the probability of an ignition source. Because risk is an integral part of system safety, it can be said that the safety process was inadequate also. See Chapter 6, Capabilities for discussions of both risk and safety capabilities. The failure of different organizations to address this issue can be put in the category of deficient interelement collaboration, which is a basic attribute of resilience. Table 4.10 summarizes some of the resilience aspects of TWA 800.

According to Federal Aviation Administration (FAA) (2008) the FAA has moved to require means to minimize flammable vapors in the fuel tanks. One

Table 4.10. Some Resilience Aspects of TWA 800

Disruption: Type B—Systemic	
Capacity	Deficient
Tolerance	Deficient
Risk Management	Deficient
Safety Processes	Deficient
Flexibility	Deficient
Inter-element collaboration	Deficient

such means is a device to inject nitrogen-enriched air (NEA) into the fuel tank cavities to purge the cavity of normal oxygen-rich air. The selection of a means will depend, in part, on the cost of implementing the change.

4.2.11 Apollo 13

Although no one died on the Apollo 13 mission April 1, 1970, this event ranks as one of the major near misses for which the skills of the crew saved the day. Chiles (2002, p. 302), Reason (1997, pp. 86–87), Leveson (1995, pp. 558–563), and Perrow (1984, pp. 271–281) all give accounts of the incident, but here is a summary of the key events: Prior to the flight, there had been a problem with an oxygen tank draining. Rather than replacing the tank, the ground crew decided to drain the tank manually and concluded that the performance of the tank would not be affected. This conclusion turned out to be incorrect because wires had been damaged in one of the tanks.

During the flight, there was an explosion in one of the oxygen tanks caused by one of the bad wires. In addition to general oxygen supply, the fuel cells on the space craft failed resulting in a loss of electrical power. In short, the four-man crew managed to survive by moving from the command module to the lunar module, which was designed to hold only two men. Needless to say, the crew managed to survive only marginally and returned to earth within an inch of their lives.

Perrow (1984, p. 274) also points to an assumption, which in Chapter 5, Culture, is called the Titanic Effect, that is to say, the assumption that the system is safe. This assumption led to a prolonged failure to detect the oxygen tank explosion for the simple reason that such an event was not considered possible in the same way that sinking by an iceberg was not considered possible for the ship Titanic. Hence, it could be said that this cultural aspect was deficient. Leveson (1995, pp. 562–563) comes to the same conclusion.

Regarding the resilience attribute of interelement collaboration, Apollo 13 stands as an example of both deficient and excellent aspects in this regard. First, the communications regarding oxygen tank draining can be regarded as deficient. However, the communications between the crew and the ground were an example of excellent interelement collaboration.

Leveson (1995, pp. 562–563) notes the following root causes: First, there were morale and budget pressures. These factors evoke the conflicting priorities paradigm discussed in Chapter 5, Culture. Next, the assumption of independence between fuel cells and oxygen points to a flawed safety analysis. Next there was a considerable error in the estimation of the probability of such an event pointing to a lack of a risk process. Finally, Leveson points to the failure to practice the use of the lunar module which had never been performed in the simulator. This neglect showed a lack of a thorough verification process.

Finally, Apollo 13 can be described as an excellent example of *adaptability* as defined in Chapter 8, Resilience Architecting. Survival of Apollo 13 can be ascribed to the primary factor of its ability to restructure itself to survive at a lower level of performance. Some resilience aspects of Apollo 13 are shown in Table 4.11.

4.2.12 Flixborough

In June 1974, an explosion in a chemical plant in Flixborough, England killed 23 people, injured 53 more, and caused $50 million in damages according to Leveson (1995, pp. 591–598), Reason (1997, p. 87), and Perrow (1984, pp. 108–112). The primary cause can be categorized as completely inadequate maintenance. However, the Board of Inquiry goes beyond simple maintenance. They also pointed to the lack of a qualified engineer on site and lack of qualified personnel performing the maintenance. The lack of maintenance and lack of expertise can be regarded as a deficiency in interelement collaboration between management and operations. So the root causes, according to the Board of Inquiry, are inadequate maintenance and lack of expertise.

Perrow (1984, pp. 108–112) points to a series of bad decisions that were made in an attempt to correct for a leak in the system. These bad decisions were associated with an attempt to get the system going as fast as possible, which refelects the conflicting priorities paradigm discussed in Chapter 5, Culture. Perrow (p. 111) characterizes this accident as the result of "a fair degree of negligence and incompetence."

Table 4.11. Some Resilience Aspects of Apollo 13

Disruption: Type B—Systemic	
Flexibility	Excellent
Interelement collaboration	Mixed (deficient to excellent)
Decision making	Excellent
Reliability	Deficient
Safety process	Deficient
Risk management	Deficient
Capacity	Excellent
Tolerance	Excellent
Culture	Deficient

Leveson (1995, pp. 596–598) points to "complacency and lack of forethought." She observed that other accidents of this type had occurred and were completely ignored. She also lists "unheeded warnings," such as the anomalies in pressure, temperature, and so on. In addition, she mentions conflicting priorities, which include, for example, cost and schedule pressures. She lists also deficiencies in organizational structure with qualified engineers and people in the right place. She says that safety activities were "superficial." These safety procedures were especially absent when changes were made, such as an increase in capacity. All in all, Leveson's observations amount to serious cultural deficiencies as discussed in Chapter 5, Culture.

Table 4.12 summarizes some of the resilience aspects of Flixborough.

4.2.13 American Flight 191, Chicago O'Hare

On May 25, 1979, an American Airlines DC-10 crashed killing all 271 people on board. According to Reason (1997, p. 88), the reason for the crash was that the engine fell off as a result of fatigue cracking and fractures in the engine pylon. These flaws were the result of procedures the airline had devised to remove the engines for maintenance. These procedures deviated from the manufacturer's Service Bulletin. Similar flaws were also found on the pylons of aircraft belonging to Continental Airlines. This case is also discussed by Perrow (1984, pp. 137–141).

In summary, the principal root cause of the accident was the improper maintenance reflected in the improper method of removing engines from the aircraft. According to Reason (1997, p. 88), two airlines, American and Continental, devised methods of removing engines that constituted a departure from the manufacturer's Service Bulletin. It can also be inferred from the above that there was inadequate communication across manufacturer-airline organizational lines. Chapter 7, Infrastructure, discusses the importance of accurate communication across organizational boundaries. This feature alone constitutes the deficiency in interelement collaboration, which is a key attribute of resilience.

Table 4.12. Some Resilience Aspects of Flixborough

Disruption: Type B—Systemic	
Maintenance	Deficient
Decisions	Deficient
Expertise	Deficient
Capacity	Deficient
Flexibility	Deficient
Tolerance	Deficient
Interelement collaboration	Deficient
Culture	Deficient

Figure 4.1 illustrates the obstacles that organizational and contractual boundaries create. This figure shows the case of an aircraft manufacturer who has to provide maintenance requirements and communicate with the maintenance organization on a regular basis. To do this, the manufacturer has to go through two layers of organizational and contractual boundaries, both of which present communication hurdles. It is these hurdles that the integrated infrastructure of Chapter 7 attempts to overcome. Table 4.13 summarizes some of the resilience aspects of American Flight 191.

In addition to the lack of adequate maintenance, Perrow (1984, p. 138) quotes from the NTSB report on the subject of multiple events all caused by the loss of the engine: (1) retraction of the leading edge slats, (2) loss of the slat disagreement warning system, and (3) loss of the stall warning system. So, although the loss of the engine leads back to the lack of maintenance, there were multiple repercussions of the engine loss.

4.2.14 Japan Air Lines JL 123, Mount Osutaka

On August 12, 1986, a Japan Airlines 747 crashed into Mount Osutaka killing all on board. This was the largest single aircraft disaster to this date. According to Reason (1997), the root cause of this accident was a botched repair job on the fuselage some years earlier resulting in a fatigue failure. Hence, this was another maintenance failure. See the Section 6.4.2.13 *Maintenance* of Chapter 6, Capabilities. Reason also points to a repair problem that made maintenance more difficult, namely, that portions of the fuselage were made more difficult to see.

It is not accurate to assume that a deficiency in repair is the responsibility of a single organization, for example, the maintenance department. This type of accident is an earmark of a lack of oversight and communications by management and therefore is a lack of interelement collaboration, which is a key aspect of resilience, as described in Chapter 8, Resilience Architecting.

Table 4.14 summarizes some of the resilience aspects of the Japan Airlines JL 123 accident.

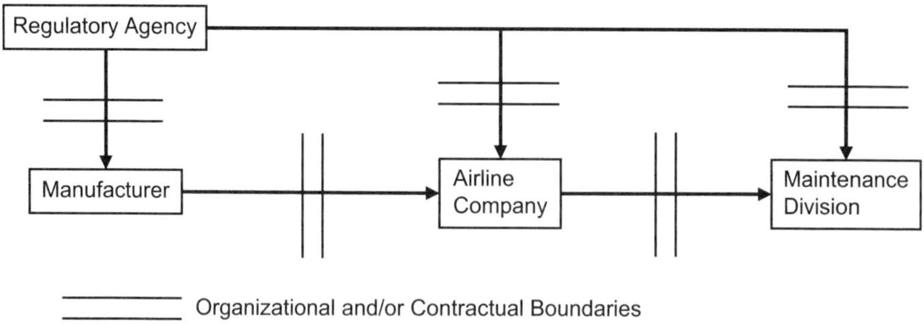

Figure 4.1. Organizational and contractual boundaries.

Table 4.13. Some Resilience Aspects of American Flight 191

Disruption: Type B—Systemic	
Maintenance	Deficient
Cross-scale interactions	Deficient
Capacity	Deficient
Flexibility	Deficient
Tolerance	Deficient
Inter-element collaboration	Deficient

Table 4.14. Some Resilience Aspects of Japan Airlines JL 123

Disruption: Type B—Systemic	
Maintenance	Deficient
Visibility	Deficient
Capacity	Deficient
Tolerance	Deficient
Flexibility	Deficient
Interelement collaboration	Deficient

4.2.15 Phillips 66 Accident

According to Reason (1997, p. 90), on October 23, 1989, an explosion in the Phillips 66 Petroleum Company in Houston, Texas killed 23 workers. The explosion resulted from the release of vapors that escaped as a result of a hose that had been fitted in the reverse position because the two ends were identical. Although the root cause of this accident was a faulty maintenance system, it is clear that the prevention of this accident would have required an attention to detail that would normally not have seemed necessary. The Section 6.4.2.13 *Maintenance* of Chapter 6, Capabilities, discusses the maintenance aspects. Chapter 9, Governance, discusses the importance of the attention to detail. Table 4.15 provides a summary of some of the resilience aspects of the Phillips 66 accident.

Similar to other case histories in this chapter, such a maintenance failure is also a failure of interelement collaboration between management and the maintenance organization on the issue of maintenance.

4.2.16 Seveso

On July 9, 1976, a chain of events began that resulted in the release of dioxin from the Icmesa Chemical Company near the town of Seveso in Italy. The dioxin resulted in the deaths of animals in the area and the hospitalization of many people. The event that initiated the disaster was a rise in temperature in the plant. Although other similar cases had occurred, these cases were

Table 4.15. Some Resilience Aspects of Phillips 66 Accident

Disruption: Type B—Systemic	
Maintenance	Deficient
Governance	Deficient
Capacity	Deficient
Flexibility	Deficient
Tolerance	Deficient
Interelement collaboration	Deficient

unknown or discounted by the plant managers. According to Leveson (1995, pp. 584–591) the following are the causal factors.

There was a lack of risk management. There was not serious consideration of risk, and a mood of complacency prevailed.

Changes were constantly being made to the plant with little or no approval or review. This deficiency reflects a basic breakdown in fundamental management processes. Hale et al. (2006) point out that the British Nuclear Installations Inspectorate has to approve all organizational changes before they are implemented. This fact is significant because, according to Weick and Sutcliffe (2001), the nuclear industry, in general, has the highest standards for safety.

The events that followed the accident showed that the parent company Givaudan had no interest in making the plant safer or addressing the problem. This deficiency showed that a culture of conflicting priorities existed. Flin (2006, p. 227), for example, states that "pressures can influence workforce behavior in a manner that drives the organization too close to its risk boundaries." This effect was certainly at work at Seveso. Chapter 5, Culture, discusses the conflicting priorities paradigm, which is detrimental to system resilience.

There was an absence of adequate safety processes. It was concluded that no real safety analysis had been done on the plant nor had any significant safety measures been put into place.

All of the above factors point to a deficiency in the interelement collaboration among Icmesa organizations in Seveso. Table 4.16 shows some of the resilience aspect of Seveso.

4.2.17 Windscale

In 1952, there was an unpredicted release of energy and radioactive material into the water around the Windscale nuclear reactor on the Cumberland coast of England. Although, according to Leveson (1995, pp. 616–619), the exact cause of the release is unknown, it is generally agreed that it had to do with the limited knowledge of nuclear reactors at that time. It is estimated that approximately 20 people may have died from thyroid poisoning from the incident.

In general, according to Leveson, the causes can be narrowed down to well-known ones, namely, complacency, lack of expertise, and a lack of a serious

Table 4.16. Some Resilience Aspects of Seveso

Disruption: Type B—Systemic	
Management oversight	Deficient
Risk management	Deficient
Change management	Deficient
Culture (conflicting priorities and complacency)	Deficient
Safety processes	Deficient
Capacity	Deficient
Flexibility	Deficient
Tolerance	Deficient
Interelement collaboration	Deficient

Table 4.17. Some Resilience Aspects of Windscale

Disruption: Type B—Systemic	
Risk Management	Deficient
Culture (Complacency)	Deficient
Expertise	Deficient
Capacity	Deficient
Flexibility	Deficient
Tolerance	Deficient
Interelement collaboration	Deficient

risk management capability. Both of these can be considered a part of a deficiency in interelement collaboration between management and other organizations. See Section 6.3.4 *Risk Management* in Chapter 6, Capabilities. Table 4.17 summarizes some of the resilience aspects of Windscale.

4.2.18 King's Cross Underground Station

In 1987, a fire broke out on an escalator in the King's Cross Underground Station in London resulting in the deaths of 31 people and injuring many more. A match or cigarette dropped by a passenger ignited the waste in the wooden escalator. According to both Reason (1997, pp. 160–161) and Leveson (1995, p. 65), the root causes were pretty clear.

First of all, this was a classic case of lax regulation. For example, the Railway Inspectorate had a narrow view of their responsibilities. Their view was that they were responsible for the operations of trains, not fires in escalators. As a result of this view, they halted routine inspections of the stations. In addition, the chief of the inquiry asserted that there was too "cozy" a relationship between the Inspectorate and London Underground. In resilience terms, this factor can be considered evidence of a lack of interelement collaboration between London Underground and the Inspectorate, which is a key resilience attribute.

Second, there was a complete misunderstanding of risk. The risk that the London Underground focused on was the risk of train collision. This fact underscores the inadequacy of conventional risk management, namely, to focus on the high-consequence and high-likelihood events that happen frequently rather than trying to identify the low-probability events that may have catastrophic consequences. This aspect calls for a deeper attention to risk as discussed in Chapter 6, Capabilities.

Leveson (1995, p. 65) focuses on one cultural aspect, namely, the assumption that the system was safe. It was assumed that fires occurred frequently and that they could be dealt with on the site. This deficiency evokes the Titanic Effect discussed in Chapter 5, Culture.

In summary, the primary root causes were as follows: First, there was lax regulation as reflected the "cozy" relation between the regulators and the Underground authority, as described by Reason (1997, p. 161). Second, there was inadequate risk management. The deficiency here, as described above, was the failure to identify the important risks correctly.

Table 4.18 provides a summary of the resilience aspects of the King's Cross Underground incident.

4.2.19 Mars Polar Lander

As discussed by Leveson (2002, pp. 30, 47, 64, 147, 249), in the year 2000 NASA sent a vehicle to Mars with the intent of landing it near one of the Mars poles. When the vehicle got within 40 meters of the surface, the vehicle struts (Agent A) deployed getting ready for the landing. The software (Agent B) was programmed to shut down the engines when the vehicle touched the surface. Unfortunately, the software interpreted the strut vibration as a sign of landing. This case is also discussed by Chiles (2002, pp. 112, 312).

So the engines were shut down prematurely, which resulted in a catastrophic landing. In short, the software did exactly what it was supposed to do and the

Table 4.18. Some Resilience Aspects of King's Cross Underground

Disruption: Type B—Systemic	
Management oversight	Deficient
Risk Management	Deficient
Regulatory oversight	Deficient
Capacity	Deficient
Flexibility	Deficient
Tolerance	Deficient
Interelement collaboration	Deficient
Culture	Deficient

struts did exactly what they were supposed to do. However, the interaction was not predicted. This is an example of a mechanical agent, the struts, interacting with a software agent, referred to in Chapter 3, Disruptions, as a case of Type B, or systemic disruption, because the interaction of the two agents could have been predicted if detailed analysis and simulation had been performed. It is also a case of two-agent disruptions as discussed in Chapter 3.

In this case, the interaction between the two system elements can be considered to be a lack of interelement collaboration, which is a key resilience attribute. However, this deficiency can be considered to be both technical—that is, between the system elements—and organizational—between the two technical groups involved.

In addition, it must be remembered, however, that neither the struts nor the software failed. The struts deployed exactly as they were designed to. The software detected the vibration exactly as it was designed to. The failure resulted from the *interaction* between two perfectly operating system elements. Leveson (2002), for example, points out that the role of perfectly operating system components is usually neglected in traditional system safety analysis.

Another conclusion is that the *anticipation* aspect was deficient. That is to say, the system was depending on a single not-too-dependable method of determining its proximity to the surface of Mars. Some other method, such as a radar altimeter, might have been a better choice.

A final conclusion on the Mars Polar Lander failure is that it cannot be considered a solely technical failure. The interelement collaboration between the engineering departments responsible for both the software and the vehicle struts was also deficient (see Table 4.19).

4.2.20 Nagoya

As discussed by Zarboutis and Wright (2006), in 1994 an Airbus A300 was approaching the airport at Nagoya, Japan. The pilot inadvertently put the aircraft in the go-around mode with the lever in the go-around position. So the aircraft flight control system (Agent A) continued to attempt to go around while the pilot (Agent B) attempted to land. Both systems lost, and all occupants of the aircraft perished. This is an example of a human agent, the pilot, interacting with a software agent, the flight control system, referred to in Chapter 3,

Table 4.19. Some Resilience Aspects of Mars Polar Lander

Disruption: Type B—Systemic	
Tolerance (hidden interactions)	Deficient
Anticipation	Deficient
Capacity	Deficient
Flexibility	Deficient
Interelement collaboration	Deficient

Disruptions, as a case of an unpredicted Type B, or systemic disruption, because the pilot's action was random and could not have been predicted by analysis or simulation. This is a clear case of a deficiency in interelement collaboration—that is, between the pilot and the flight control system.

Table 4.20 summarizes some selected resilience aspects of the Nagoya accident.

4.2.21 Sioux City–1989

On July 19, 1989, a DC-10 approaching Sioux City, Iowa lost all control of control surfaces on the aircraft. According to the Aviation Safety Network (1989), the pilot with the assistance of an off-duty mechanic managed to land the aircraft using the engine throttles. The original failure was caused by deficient maintenance on the engines. Despite this effort, there were 111 fatalities out of a total of 296 occupants. Captain Haynes (1991) also credits excellent communications with the ground for preventing further loss of life. The resilience attribute of interelement collaboration can be considered mixed. First, the faulty maintenance was an example of deficient intercomponent collaboration between maintenance and management. Second, pilot communications with the ground are an example of good inter-component collaboration. The pilot's ingenuity is an example of superior flexibility, which is another resilience attribute. Reason (1997, p. 79) comments that this incident "was a remarkable feat of airmanship and crew resource management." Table 4.21 presents some of the resilience aspects of the Sioux City accident.

Table 4.20. Some Resilience Aspects of the Nagoya Accident

Disruption: Type B—Systemic	
Tolerance (hidden interactions)	Deficient
Human-machine knowledge, control, predictability	Deficient
Capacity	Deficient
Flexibility	Deficient
Interelement collaboration	Deficient

Table 4.21. Some Resilience Aspects of the Sioux City Accident

Disruption: Type B—Systemic	
Maintenance	Deficient
Interelement collaboration	Mixed (maintenance and pilot)
Tolerance through engine control	Good
Capacity	Adequate
Flexibility	Excellent

Haynes also pointed to ongoing studies to consider the possibility of flight control systems that would control the aircraft using the propulsion control. Although this research looked promising, no action has been taken to mandate propulsion control. However, in the end the National Transportation Safety Boaed (NTSB) (1990) did not recommend automated propulsion control but rather focused on a number of preventive measures including design and maintenance procedure changes.

4.2.22 Tacoma Narrows Bridge

The Tacoma Narrows bridge collapse of 1940 is an example of a case in which the designers were completely unaware of the physical phenomena that were responsible for the accident. The root cause of this accident was that an unstable boundary layer vortex developed under the bridge causing a pressure differential on the surface of the bridge. When the bridge rose up under the force of the pressure, the vortex disappeared, causing the pressure differential to disappear and the bridge to fall. As the bridge descended, the pressure differential reappeared causing the bridge to rise again, which created a repetitive oscillation of the bridge that eventually caused it to collapse. In the final moments, the oscillations were in a torsional mode.

In addition to the aerodynamic effect, the bridge designers had also made shortcuts in structural strength, so it can be said that capacity was deficient. It might also be possible to say that if the aerodynamic effects had not been there, then the bridge might have lasted many years. However, the combination of the two factors was sufficient for failure.

In the context of Chapter 3, Disruptions, this is a case of Type A, or external disruption. This is also a case of a one-agent disruption because only one system was involved, namely, the bridge. It can also be said that this disruption was unpredictable because the bridge designers were unfamiliar with the phenomenon of a turbulent boundary layer.

The only positive thing to say about the Tacoma Narrows bridge is that it did not collapse immediately. In resilience terminology, it can be said to experience a graceful degradation so that no actual fatalities occurred. At the same time it can be said that although the detection of drift to catastrophe was good, local decision making was deficient. There is no evidence that steps were taken to keep people and cars off the bridge before it collapsed. Hence, interelement collaboration was deficient also. Table 4.22 summarizes some of the resilience aspects of the Tacoma Narrows bridge disaster.

4.2.23 New York Power Restoration

Following the attacks on the twin towers on September 11, 2001, the city of New York was without electrical power for the lower portion of the island of Manhattan. Amazingly, according to Mendoça and Wallace (2006), power was restored within 5 hours after the attack. The system of interest in this case

Table 4.22. Some Resilience Aspects of Tacoma Narrows Bridge

Disruption: Type A—Input	
Capacity	Deficient
Expertise	Deficient
Tolerance	Adequate
Detection of drift	Adequate
Decision making	Deficient
Interelement collaboration	Deficient

Table 4.23. Some Resilience Aspects of New York Power Restoration

Disruption: Type A—Input	
Decision making	Excellent
Reserve capacity	Excellent
Cross-scale connectivity	Mixed
Flexibility	Excellent
Tolerance	Good

is the power system itself with some associated nodes. The system consists of all the hardware, particularly a large supply of portable generators, and the organizational entities: the power company, the U.S. Army, the New York police, and the New York Fire Department.

This case represents a major example of the attribute of flexibility to be discussed in Chapter 8, Resilience Architecting. The flexible aspect was the supply of portable generators and the action by the power company to form an instant suborganization to deploy and manage the generators. The interelement collaboration characteristic was the cooperation among the four agencies mentioned above. The one interelement characteristic not present was the ability to get the government of New Jersey to allow fuel trucks through the tunnels connecting New York and New Jersey. This problem was eventually resolved through the intervention of the Federal Emergency Management Agency (FEMA). The resilience aspects are summarized in Table 4.23.

New York Power Restoration can be considered a case of a Type A, or external disruption, as described in Chapter 3, Disruptions, because it was initiated by humans whose actions are unpredictable.

4.2.24 ValuJet

The ValuJet crash of 1996 is a classic case of lax regulation combined with poor communication among the elements of the infrastructure nodes. Chapter 7, Infrastructure, defines an infrastructure as the set of organizational nodes responsible for the success of a system, in this, case a commercial aircraft. Flawless communication of information is required at all levels of the infrastructure to assure success, particularly safety critical information. Chapter 5, Culture, highlights the difficulties in achieving this communication because of cultural factors that result in a focus on one organization without attention to cross-organizational communication. Chapter 8, Resilience Architecting calls this factor an interelement collaboration and shows how this factor is critical to resilience to a major disruption, such as the ValuJet experience.

The essence of the accident was, first, according to Chiles (2002, pp. 150–155, 310), the FAA had failed to report fires in expired oxygen canisters that had been happening for several years in several airlines. However, worse that that, the transportation and storage of the expired oxygen canisters were a quagmire of bureaucracy and sloppiness. The independent contractor responsible for loading the canisters was under cost and schedule pressure and made only a cursory review of safety procedures. It was concluded that during flight the canister retaining pins came loose because of the poor packing and then ignited, which caused the fire and ultimate loss of the aircraft.

Although there were several contributing factors to this accident, as there usually are, a key one is the lack of communications across the infrastructure. Although communications between the manufacturer and the airline may have been as expected, the added communication between the airline and the contractor resulted in a deterioration of information. This factor, combined with the lax regulation and cost and schedule pressures, all added up to one result: disaster. Table 4.24 provides some of the resilience aspects of the ValuJet accident.

4.2.25 Katrina

In the fall of 2005, Hurricane Katrina hit the city of New Orleans, Louisiana with devastating results. It was a classic case of a lack of resilience. But what was the system that lacked resilience? First, there were the physical elements: the Mississippi river levees, the transportation system, housing, communications equipment, and so on. However, more important than any of these was the organizational infrastructure that had responsibility for assuring the safety of the people. This included the police, the U.S. Army Corps of Engineers, the local government, and FEMA, to name a few. In short, there was almost a complete absence of communication and resources among agencies.

Table 4.24. Some Resilience Aspects of the ValuJet Accident

Disruption: Type B—Systemic	
Inter-element collaboration	Deficient
Safety	Deficient
Culture	Deficient
Capacity	Deficient
Flexibility	Deficient
Tolerance	Deficient

The failure of FEMA to coordinate the activities among all other organizations is the conventional conclusion of Hurricane Katrina. This is the view of Westrum (2006b). However, there is another view, namely that a resilient system should not be dependent on a single organization like FEMA. Their view is that if the system were resilient, then local agencies would have a robust communication system linking each other and that resources would be shared among them. In the terms of Chapter 8, Resilience Architecting, there was an absence of interelement collaboration, a key attribute of resilience. Stephan (2007) cites two main reasons that communications were so poor: First, the state had decided not to pay the access fee for the federally provided communication system. Second, the local communication systems could not talk to each other because of a lack of compatibility of the equipment. In communications terminology, this is called a lack of *interoperability*.

Another aspect of resilience that is completely technical is the lack of *capacity* in the levees that were supposed to protect the cities. With both a weak infrastructure and inadequate physical capacity, the city had no tolerance for this storm even though storms of this magnitude were not unknown in this area. The civil infrastructure system was also deficient in flexibility because it did not have alternative mode of transportation or shelter.

In summary, Hurricane Katrina is a classic case of a lack of resilience from both an infrastructure point of view and a technical point of view. Table 4.25 summarizes the resilience aspects of the civil infrastructure system affected by the Katrina hurricane.

As Table 4.25 shows, Hurricane Katrina can be considered to be both the result of a Type A, input and a Type B, systemic disruption. On the surface, Type A would seem to be the obvious classification because the disaster resulted from a natural phenomenon, a hurricane. However, the internal problems were so severe, for example the neglect of the levees, that this disaster should also fall into the Type B category as well.

4.2.26 Mars Climate Orbiter

As described by Chiles (2002, pp. 111–112, 311), in 2002 a NASA mission to study the climate of Mars failed when a flight control computer transmitted its data in English, rather than metric, units. The root cause of this failure was

Table 4.25. Some Resilience Aspects of the Katrina Civil Infrastructure System

Disruption: Type A—Input and Type B—Systemic	
Decision making	Deficient
Interelement collaboration	Deficient
Capacity	Deficient
Flexibility	Deficient
Tolerance	Deficient

generally considered to be a failure to use the metric system. However, as described in Chapter 10, Measuring Resilience, the failure to use the metric system could only be a symptom, not a systemic cause. The systemic cause was the failure to employ the standard interface management, which is a standard process in the aerospace domain. See Section 6.4.2.4 Interface Management in Chapter 6, Capabilities. In resilience terms, this lack of capability between two system elements is known as a deficiency in interelement collaboration both at the technical and managerial level.

An interface management system would have enhanced the adaptability of the system as described in Chapter 8, Resilience Architecting. That is, with an interface management process in place, it would not have mattered what units were used. But, it should be asked, why would such a basic process as interface management be neglected on such a complex system as Mars Polar Lander? It can only be concluded that technical management failed in this respect. The resilience aspects of the Mars Climate Orbiter are summarized in Table 4.26.

This incident can also be described as a case of Type B, or systemic disruption, as described in Chapter 3, Disruptions.

4.2.27 Comet

On two separate occasions, January and April of 1954, according to Chiles (2002, p. 299) two British airliners, the Comet, exploded over the Mediterranean. The Comet was the world's first commercial jet airliner. The cause was attributed to cracks in the fuselage, that is, metal fatigue. There were a total of 56 fatalities including passengers and crew.

The phenomenon of metal fatigue was well known in the engineering community long before the Comet. Therefore, one might ask, why was it not anticipated on the Comet? The simple answer is that it was a remote phenomenon and that jet propulsion was a new technology about which peripheral consequences, such as metal fatigue, little was known.

From a technical point of view, these accidents could be categorized as a deficiency in capacity, that is, the capacity to sustain the disruption. However, because there were no alternative ways to survive, the system could be said to lack flexibility and tolerance as well. From a managerial point of view, the

Table 4.26. Some Resilience Aspects of Mars Climate Orbiter

Disruption: Type B—Systemic	
Analytic methods (interface management)	Deficient
Technical management	Deficient
Capacity	Deficient
Flexibility	Deficient
Tolerance	Deficient
Interelement collaboration	Deficient

failure to consult experts could be considered a deficiency in interelement collaboration.

Hence, Comet can be considered a case of Type B, or systemic, disruption, as discussed in Chapter 3, Disruptions, because the causes were known and could have been predicted with sufficient analysis. This incident is similar to the Tacoma Narrows bridge disaster discussed earlier in this chapter in this respect. With respect to metal fatigue, another aspect is of importance. It was a practice at this time to stamp part numbers on the individual aircraft components using marks that left depressions in the metal. This practice was considered as a contributing factor in the accidents. Since that time, the standard practice is to mark the parts using ink.

The resilience aspects of the Comet are summarized in Table 4.27.

4.2.28 Concorde

On July 25, 2000, the Concorde supersonic transport crashed near the Charles De Gaulle airport in France killing 100 people including crew according to the Bureau Enquétes Accidents (BEA) report (2000) and as described by Chiles (2002, pp. 4–5, 312). The immediate cause of the accident was a band of metal that had fallen from another aircraft that had taken off only 5 minutes before the Concorde. This band of metal severed the tire of the Concorde, which ultimately ruptured the fuel tank.

This accident illustrates several points that are discussed later in this book. First of all, the band of metal had fallen from an aircraft that had not been maintained in accordance with the specifications of the manufacturer. The similarities to the ValuJet crash, discussed above, in this respect, are apparent. First, the sensitivities to maintenance issues are apparent, as discussed under 6.4.2.13 Maintenance in Chapter 6, Capabilities. Second the communications disconnects between manufacturers and suppliers continue to reduce resilience, as discussed above in the ValuJet case.

Finally, the Concorde itself demonstrates an extreme vulnerability to small problems, thus showing a severe lack of tolerance, as described in Chapter 8, Resilience Architecting. From a resilience point of view, this vulnerability was a lack of tolerance to small disruptions, namely, the band of metal. The Concorde

Table 4.27. Some Resilience Aspects of the Comet Accidents

Disruption: Type B—Systemic	
Analytical methods (interface management)	Deficient
technical management	Deficient
Capacity	Deficient
Flexibility	Deficient
Tolerance	Deficient
Interelement collaboration	Deficient

system itself lacked the capacity to survive such a disruption. Because there were no alternative ways to survive, the system could also be said to lack flexibility. Regarding interelement collaboration, this can only be considered to be deficient in a global sense, that is, the lack of knowledge between Concorde system itself and the external system that created the disruption. The resilience aspects of the Concorde aircraft are summarized in Table 4.28.

4.2.29 Jésica Santillán

In September 2003, a 17-year-old girl, Jésica Santillán, had been brought to the United States for a heart transplant. According to Kopp (1993), the heart provided by the blood donor organization was a different blood type from Jésica's. By the time another heart could be implanted with the correct blood type, it was too late and Jésica died. This case received considerable media attention.

Aside from the attention it received, there are several reasons this case is worth including in this list. Hospitals are an example of human-intensive systems in which the potential for serious mistakes is high. Okada (2003), for example, analyzes the performance-shaping factors (PSFs) that contribute to human errors in hospitals. Second, hospitals are an example of a system of systems in which several independent system elements are expected to work together to achieve a goal. In this case, the goal was saving the girl's life. Depending on the situation, the independent system elements may include the personal physician, the hospital itself, and the donor organization.

In terms of the attributes shown in Table 4.29, it is likely that capacity of the system was adequate, that is, there were sufficient numbers of personnel and equipment. In terms of flexibility, however, the system was deficient. There did

Table 4.28. Some Resilience Aspects of the Concorde Aircraft

Disruption: Type A—External	
Capacity	Deficient
Flexibility	Deficient
Tolerance	Severely deficient
Interelement collaboration	Deficient

Table 4.29. Some Resilience Aspects of the Jésica Santillán Case

Disruption: Type B—Systemic	
Capacity	Adequate
Flexibility	Deficient
Tolerance	Deficient
Oversight	Deficient
Interelement collaboration	Deficient

not seem to be a mechanism within the system to obtain a second heart. It took about 2 weeks to get another one. Of course, tolerance was deficient. An incorrect blood type in this situation will result in almost immediate death. A flexible organization could have compensated for this factor. The reports on this case focus on the lack of double checking, which is a mandated practice. This lack of double checking contributed to the lack of tolerance.

A primary deficient attribute was the lack of interelement collaboration, in this case, the communications between the hospital and the blood donor organization. We saw in Chapter 1, On Resilience, that interelement collaboration is a key attribute of resilience. Okada (2003) identifies "interfaces" as a key PSF in hospitals. That is, mistakes are most likely to happen over interfaces. That is exactly what happened between the blood donor organization and the hospital in this case. This was also a primary factor in the ValuJet case discussed earlier in this chapter.

4.2.30 Helios 522

In August 2005, a Helios Airlines Boeing 737 took off from Cyprus with a German pilot and Cypriot copilot. Because of to a malfunctioning pressurization system and an automated flight control system that worked as designed, the pilot, copilot, and passengers all died of hypoxic hypoxia.

According to Dekker and Hollnagel (2006), this is a classic case in which the system lacked the adaptability to deal with the interaction of the components that was the root cause of the accident. In Chapter 3, Disruptions, this is called a two-agent disruption. Limited English by both the pilot and copilot also contributed to the accident.

As in the case of the Mars Polar Lander, above, Helios is an example of an accident in which no components failed. Although the crew failed to enter the correct settings for the environmental control system and also failed to heed the warning signals, it was the interaction between the environmental control system and the flight control system, which operated as designed, that led to the accident. This is yet another example, as pointed out by Leveson (2002), in which a normally operating system element contributed to the accident. It was the interaction between these elements that resulted in the accident. The Helios 522 resilience aspects are summarized in Table 4.30.

Table 4.30. Some Resilience Aspects of the Helios 522 Accident

Disruption: Type B—Systemic	
Capacity	Adequate
Flexibility	Deficient
Tolerance	Deficient
Oversight	Not applicable
Inter-element collaboration	Deficient

According to Federal Aviation Administration (FAA) (2008b), the FAA has issued interim regulations to change the operating procedure on that aircraft to avoid the situation that led to the Helios 522 accident. The FAA is working with the manufacturer for a longer term solution.

4.2.31 Metrolink 111

On September 12, 2008, a Los Angeles Metrolink commuter train, number 111, collided with a Union Pacific freight train in Chatsworth, California near Los Angeles. There were 25 fatalities and many injuries.

Immediate attention fell on the train's engineer, Mr. Robert M. Sanchez, who died in the crash. According to Lopez and Hymon (2008b), Sanchez was text messaging on his cell phone immediately before the crash. It was confirmed, according to Archibold (2008), that the engineer did not apply the brakes. Morrison (2008) reports that a Metrolink spokesperson immediately placed the blame on Metrolink because it was a Metrolink engineer who was at fault. The spokesperson, Denise Terrell, did not address any systemic problems, only the actions of the engineer.

Another contributing factor being studied was the fact that the engineer was fatigued because of a "split shift," that is, working two 5-hour shifts in a day.

There was general awareness, however, that the engineer was not the only problem. A multitude of solutions were proposed, both in the California legislature and in the hearing before the NTSB. These include an advanced train stopping system described by Hymon and Dizikes (2008), a second engineer in the cab and extra video cameras described by Hymon and Connell (2008), banning of cellphones as described by Connell and Lopez (2008b), and older train stopping devices described by Lopez and Weikel (2008).

What can be concluded? First, the engineer, as ill-advised as his actions may have been, can be considered the primary disruption in the incident. In Chapter 3, Disruptions, we saw that actions such as his were disruptions from which disasters may occur. The ultimate cause of the disaster was the lack of resilience of the Metrolink system. Once again, the engineer's actions cannot be excused. However, there is a greater responsibility for this event as pointed out by Meshkati and Osborn (2008).

As observed in Table 4.31, there is almost no aspect of this system that was resilient. There were no alternative ways to avoid this accident; therefore, capacity was deficient. There were no opportunities to reconfigure the system; therefore, flexibility was deficient. There was no opportunity for gradual degradation; therefore, tolerance was deficient. There was no interelement collaboration that might have prevented this event.

Appendix B will present a more in-depth discussion of this accident and the various options that have been proposed.

Table 4.31. Some Resilience Aspects of the Metrolink 111 Accident

Disruption: Type B—Systemic	
Capacity	Deficient
Flexibility	Deficient
Tolerance	Deficient
Inter-element collaboration	Deficient

Table 4.32. Some Resilience Aspects of the US Airways 1549 Accident

Disruption: Type A—External	
Decision making	Excellent
Capacity	Limited
Flexibility	Excellent
Tolerance	Excellent
Interelement collaboration	Excellent

4.2.32 US Airways 1549

On January 15, 2009, the pilot of U.S. Airways Flight 1549 ditched his Airbus A320 into the Hudson River in New York just after taking off from La Guardia Airport. There were no fatalities among the 155 people on board. The supposed disruption was that the aircraft had struck a flock of birds, perhaps geese, which cause both engines to fail. Thus, the disruption would be a Type A, or external, disruption. The basic account of this accident is provided by McFadden (2009).

Although bird strikes are not uncommon for flying aircraft, and all jet engines are subjected to bird-strike tests, the magnitude and character of any bird strike are impossible to predict. Shutting down two engines at the same time is also rare. Thus, it is not possible to design an engine that will survive a bird strike of any size or kind. Hence, it is important to design an aircraft that is resilient to these disruptions. Hence, the system's *capacity*, as defined in Chapter 2, System Resilience and Related Concepts, and discussed more thoroughly in Chapter 8, Resilience Architecting, to survive such a disruption is limited, and other resilience attributes are necessary.

The pilot, Chesley B. Sullenberger, III, received deserved credit for his quick action to bring the plane down in water rather than return to La Guardia or fly to a nearby alternative airport. His actions were examples of the critical decision-making process for humans on the "sharp edge" of a system, as discussed in Chapter 2. These actions require both a natural ability and also simulator training in similar conditions. It is assumed that Sullenberger had both. It is also reported that Captain Sullenberger had glider training, which may have contributed to the success of the ditching.

The published reports on this accident say little about the aircraft itself, the Airbus A320. However, for the pilot to accomplish his feat, the aircraft needed certain capabilities. The first capability the aircraft needed was the ability to maintain control without the engines. On most commercial aircraft, like this one, the primary means of control is the propulsion system itself, which provides power to the hydraulic system that controls the aerodynamic control surfaces. In the case of a propulsion system loss, like this one, power is transferred to the auxiliary power unit (APU). Hence, *functional redundancy* is maintained, which is a primary characteristic of resilience.

The second aircraft capability needed was the ability to control the pitch, the vertical angle, of the fuselage while ditching. The pilot needed to maintain a high nose and a low tail so that the tail would strike the water first and slow the plane down, which it did.

The third thing that was needed was the capability of floating. On this aircraft, the pilot was able to send a command closing all openings that would cause the plane sink in the water. This characteristic contributed to its gradual degradation, or *tolerance* to disruptions.

There was *interelement collaboration* at all levels. The pilot was able to communicate with the air traffic control (ATC) system at all times and was able to send internal commands to the control system. The external communications resulted in immediate rescue by boats from the New York Fire Department, New York Police Department, the U.S. Coast Guard, and other agencies.

In short, all resilience attributes were present to help this aircraft survive and recover from a major disruption. All of these factors are summarized in Table 4.32.

4.3 SUMMARY

In summary, although the above case histories represent only a few of many possible examples of catastrophes, the associated root causes do suggest patterns of commonality regardless of the domain, commercial aircraft, nuclear power, chemical processes, and so on. This commonality also suggests that common approaches may also be applicable across domains. These common approaches are reflected in, for example, the system resilience capabilities discussed in Chapter 6, in the infrastructure aspects discussed in Chapter 8, and in the cultural aspects discussed in Chapter 5.

4.4 FURTHER EXPLORATION

1. In your studies of case histories of catastrophic events from the literature, discuss several such events and their root causes. Describe the root causes in terms of systemic causes rather than symptoms. Be sure to include both human systems and product systems in your discussion.

Chapter 5

Culture

. . . the magnitude of disasters decreases to the extent that people believe that disasters are possible and plan to prevent them or to minimize their effects.

—Leveson (1995, p. 57) paraphrasing the phenomenon known as the Titanic Effect observed by Kenneth E. F. Watt

Probably the most challenging transdisciplinary aspect of resilience is culture. This is because culture is traditionally the domain of organizational psychology and sociology, not engineering. Nevertheless, solutions are sought. Either systems need to be designed that are impervious to cultural factors, or alternatively, cultures need to be changed.

The term "culture" may refer to the individual, organizational, or national beliefs that govern our actions. Although the cultural influences on resilience are well documented, this aspect has received little attention in industry or government for several basic reasons: First, culture is not technical or quantifiable. Second, when it is considered, the solutions are simplistic and not based on any sound analysis or research.

When a cultural belief is identified, it is called a *paradigm* or mindset. Many paradigms have been identified that have a negative effect on resilience; that is, they contribute directly to accidents. Following are some examples of some negative paradigms that have been identified. These paradigms are transdisciplinary because they reflect both technical and managerial processes, both of which are critical to resilience. This list is extracted from a previous paper by Jackson et al. (2006).

This chapter addresses challenges that may be obstacles to system resilience. These challenges include responsibilities in safety, management, contracts, cost, and schedule. This chapter discusses some approaches to achieving

Architecting Resilient Systems: Accident Avoidance and Survival and Recovery from Disruptions.
By Scott Jackson

breakthrough by changing behavior. Some people may believe that cultural change can be achieved by lectures, intensive training, or simply by good engineering processes. Experts in the field of organizational development, management, and leadership suggest that these methods are not sufficient. This chapter demonstrates the true transdisciplinary nature of system resilience by showing that such diverse fields as organizational psychology are essential to the success of a system.

5.1 THE CULTURE ELEMENT

Most system resilience models include a culture element; that is, they recognize the importance of beliefs or paradigms of the people working in that industry or domain. These paradigms can often be an obstacle to the accomplishment of system resilience.

While many people might consider negative paradigms as *irrational*, it can be argued that all paradigms are rational in the sense that they reflect the cultural environment of the opinion holder. On the other hand, some paradigms can be seen to lead to higher risk opinions than others. It is these opinions on which this chapter focuses.

Traditionally, organizations tend to deal with these paradigms in two ways: First, they depend to a great extent on executive influence, that is, leadership. They assume that if the organization's leaders are good role models, then this influence will filter down through the organization. Second, they depend on training to disseminate the message of the leaders. It is the thesis of this chapter that these methods have been shown to be inadequate in the face of highly complex and vulnerable systems and that the science of organizational psychology can show us the way to better approaches.

5.2 POSITIVE AND NEGATIVE PARADIGMS

Vaughn (1996) points out that paradigms can be either positive or negative, in other words, beneficial or detrimental to the project at hand. She begins by discussing the scientific paradigm which she characterizes as mostly positive. She says (p. 196) that "it [the scientific paradigm] is a world view based on accepted scientific achievement, which embodies procedures for inquiring about the world, categories for which these observations are fitted, and a technology that includes beliefs about cause-effect relationships and standards of practice and behavior." She adds (p. 402) that "most science and technology is done in social organized settings that can hardly be described as neutral." In short, Vaughn is saying that the social environment may be in conflict with the scientific paradigm and indeed may corrupt it, which is what was the case for the Challenger mission. She describes this environment (p. 394) as "NASA's institutional history of competition and scarcity."

The primary social paradigm that Vaughn discusses throughout her book is what she calls the "normalization of technical deviation" (p. 113). This is the paradigm that conflicts most strikingly with the scientific paradigm. This paradigm reflected itself in what NASA calls "acceptable risks." She says (p. 82) that "flying with acceptable risks was normative at NASA." This practice will be discussed later in this chapter as part of the *risk denial* paradigm.

5.3 THE CULTURAL END STATE

Both Weick and Sutcliffe (2001) and Reason (1997) do an excellent job of defining the end state of the culture of any organization or the entire infrastructure. The premise is here that when the cultural paradigms discussed later in this chapter are conquered, this cultural end state will be a natural result. There is much agreement between the two sources. Following is a summary of their conclusions.

5.3.1 Preoccupation With Failure

Weick and Sutcliffe (2001) envision a culture that is mindful, that is, it pays attention to details, especially those details that may result in failure. Mindfulness applies not only to managers but also to all members of the organization. Chapter 9, Governance, discusses those small defects that are intellectual as well as physical that may result in failure. The detailed cross checks discussed in that chapter are an institutional support to the culture of mindfulness.

Although most people might consider fear of failure to be a bad thing, D'Antonio (2002) says that it is a good thing. D'Antonio says that people's attention to the possibility of failure increases greatly after a major accident. So, how can this fear be instilled *before* a major accident occurs? This is the challenge. The approaches discussed later in this chapter provide a roadmap.

5.3.2 Reluctance to Simplify Interpretations

Everyone reading this book has probably witnessed cases in which outspoken people in an organization were marginalized or ostracized. These people find problems that are often ignored or simplified. In the future culture described by Weick and Sutcliffe (2001) these people would be both tolerated and listened to. The future culture would recognize the complications and subtleties that may lead to failure.

5.3.3 Sensitivity to Operations and a Reporting Culture

Being sensitive to small problems that may grow into big problems is the essence of this quality of system resilience. Leveson et al. (1995), for example, points out that there is a correlation between defects and the risk of major catastrophes. This risk is even stronger when the defects are determined to be systemic in nature. Section 6.3.7, under Corrective Actions, for example, points out that when defects are found, it is the responsibility of the investigating team and the independent review team to determine whether the defect is an anomaly or systemic. Reason (1997) says that near misses, his term for defects, "provide quantitative insights into how small defective failures can line up to create large disasters." The focus of this chapter is the cultural view of these defects and whether they receive adequate attention. Chapter 10, Measuring Resilience will discuss the quantitative predictability of these defects.

Reason supports the above conclusions but points to the importance of a reporting culture. The point is that if defects are not properly identified, then it will be impossible to infer the possibility of a catastrophic failure. Achieving the reporting culture requires two things: first, a reporting culture, and second, a robust corrective action system with independent reviews as described in Chapter 9, Governance.

5.3.4 Commitment to Resilience and a Learning Culture

Weick and Sutcliffe (2001) characterize a commitment to resilience as a culture willing to learn from its mistakes and build on them. Reason calls this a learning culture.

5.3.5 Deference to Expertise and a Flexible Culture

Weick and Sutcliffe (2001) characterize a system resilience culture as one that defers to expertise. Reason describes this culture as flexible; that is, it is a culture that depends on its experts to decide what way to go next. Remember that expertise pertains to the capabilities listed in Chapter 6, Capabilities.

According to Freedman (2006, p. 38) the NASA administrator, Michael Griffin, overrode the chief engineer and top safety officer and ordered the launch of the space shuttle *Discovery*. Fortunately for Griffin and NASA, the mission was a success. This decision might be considered a gamble that paid off.

5.3.6 Just Culture

Reason discusses one issue that cuts across all the previous characteristics of an ideal cultural end state, that is, the issue of justice. In the just culture, a person who makes a mistake would not get an automatic amnesty. The rewards and punishments would be made clear. But, at the same time, a person would not receive punishment for simply reporting a problem or error.

5.4 PARADIGMS

This section will enumerate paradigms extracted from the literature and from the author's experiences. In the author's judgment, these paradigms are *potential* obstacles. That is, the author wishes to refrain from making judgments regarding which are the true paradigms because, in our opinion, the reader should discover for himself or herself. In some cases, the paradigms have no basis in science or mathematics. In such cases, it is necessary to point out these flaws.

5.4.1 Paradigms versus Heuristics

Later in Chapter 8, Resilience Architecting, we will study a collection of guidelines called heuristics, that is, lessons from experience that have been found to *improve* resilience of a system. The paradigms in this chapter have been found, however, to *reduce* resilience. Hence, heuristics of Chapter 8 can be considered to be a positive counterpart to the negative effects of the paradigms in this chapter.

5.5 COMMON AND ALTERNATIVE PARADIGMS

The paradigms listed below have been observed either by the researchers cited in the individual paragraphs or by the authors of this book and the authors of various references in their professional lives.

5.5.1 The High-Level Problem Paradigm

It is the belief of some people that they should focus only on large problems and ignore small problems. Their job, in their opinion, is to address large problems. This idea, although not totally irrational, is counter to historical evidence. Most disasters involve very small, and sometimes imperceptible, aspects of a system. The O-rings (on Challenger) and the tiles (on Columbia) were very small components.

Weick and Sutcliffe (2001) point to the major disasters that have their roots in small problems. It is their contention that all industries, especially those with the potential for high-consequence events, should make aircraft carriers and nuclear power plants the model for management. These industries, they say, have a relatively good record for one reason: They focus on small problems.

Other authors, for example, Reason (1997) and Leveson (1995), point out that large problems, such as major disasters, are almost always preceded by small problems and near misses that could have been used as indicators of impending large problems.

Dekker (2006) points to the distance between how management sees processes being performed and how they actually are performed.

Weick and Sutcliffe (2001) refer to the attention to detail as mindfulness. One characteristic, they say, of mindfulness is "a preoccupation with failure." This idea raises the question, is a preoccupation with failure really different from a preoccupation with success? After all, one is just the reverse of the other. Yes, it should be said, there is a critical difference. The difference is in the focus. If you focus on success, then you lose the focus on failure. As Weick and Sutcliffe say, it is the focus on failure that is the key to system resilience.

Citing a study comparing National Aeronautics and Space Administration (NASA) with the U.S. Navy, the latter being more successful, Wald (2003) notes that the Navy has a practice of documenting all day-to-day decisions and sending them to the admiral in charge. In contrast, Wald notes that the NASA administrator, Mr. O'Keefe, has shown himself to be unaware of details of the shuttle program.

Wald and Schwartz (2003, p. A9) provide another example of the high-level problem paradigm when they note that Edward R. Tufte, professor emeritus at Yale, observed of the Columbia disaster, "In a real sense, the real fault of upper management is that they didn't look beneath the optimistic surface of the reports of their subordinates."

5.5.2 The Titanic Effect

At the beginning of this chapter Leveson (1995, p. 57) describes the psychological phenomenon of the Titanic Effect coined by Kenneth E. F. Watt. The Titanic Effect is, in essence, a belief that a system is safe when it is not. Watt was referring, of course, to the ill-fated ocean liner that struck an iceberg in 1912 with devastating results. Leveson was also referring to the overconfidence of the builders who declared that the Titanic was an absolutely safe ship. The Titanic Effect was also present in the Challenger and Columbia disasters.

Another manifestation of the Titanic Effect is the misguided belief that risk decreases over time. Leveson (1995) points out that risk may actually increase over time for various reasons: caution may wane, the system ages, and the system changes.

Modern systems are not immune to this phenomenon. This paradigm becomes more important as complacency sets in, especially after a period of successful operation by the system.

The Titanic Effect also is often present in start-up ventures with more optimism than caution. The following two examples have not had any major catastrophes, but their public statements reflect this optimism.

In a television interview, the entrepreneur Richard Branson (2007), who is sponsoring a private space venture called SpaceShipOne, said that all he had to do was have a dream and then leave it up to the engineers to accomplish it. On the face of it, Mr. Branson seems to have minimized the importance of the management of the program and the responsibility that management has for implementing resilience. To be fair to Mr. Branson, he was perhaps including management in the use of the term "engineers."

Second, Schwartz (2008) relates how Jeff Bezos has launched the development of a space vehicle called the Blue Origin (see Figure 5.1). Mr. Bezos emphasizes the goal of exploring the solar system more cheaply and using experienced engineers to do it. To his credit, Mr. Bezos does advocate an incremental step-by-step approach. Programs such as Blue Origin may be tests of whether a private venture can architect, develop, launch, and operate a more resilience space system than a public agency.

Figure 5.1. The Blue Origin space vehicle.
Source: Blue Origin (2007). Retrieved on March 30, 2009, from http://public.blueorigin.com/letter.htm, [crop]. Reprinted with permission.

5.5.3 The Informal Cost Analysis Paradigm

Often in organizations you will find quick, informal cost analyses done. The result of these analyses is usually a statement like "We can't afford to verify everything." The implication is that some requirements will go unverified. This statement is contrary to the principle that all requirements must be verified.

This principle does not imply that costly tests need to be performed. Many requirements may be verified by analyses such as similarity.

Another common statement is that "we can't afford to make the system safer." Although the Federal Aviation Administration (FAA) does include the value of human life in their analyses, as discussed in Chapter 11, design decisions are rarely made with this level of analysis.

There is an implicit assumption in this paradigm that a standard depth of analysis is associated with all systems in all domains. It is more logical to assume that the depth of analysis reflected by the capabilities of Chapter 6, Capabilities, will depend on many factors. The probability of a disruption is only one such factor. Other factors discussed in Chapter 12, Implementation, include the consideration of the cost of the capability. Chapter 12 shows that some capabilities can be implemented at little or no cost. Another consideration is the consequence level and risk of the system in mind. Nuclear power plants, aircraft carriers, and space vehicles, for example, deserve more attention than other systems. Another gauge of the extent of implementation is the benefit-cost ratio, which can be determined more easily in some domains than others. Chapter 11, Cost, shows that the commercial aircraft domain has had a measure of success with this determination.

Another factor in this paradigm is the difficulty most people have with understanding probabilities. The most striking example of this difficulty is the common perception of multiple events and their association with the Law of Large Numbers, as discussed in Chapter 3, Disruptions. In short, even though the probability of the undesirable interaction among elements of a system may seem to be infinitesimally small, the probability of any two elements *of the group* interacting may be significant.

5.5.4 The Independent Supplier Paradigm

This paradigm is the belief that suppliers can figure out what to build without giving requirements to the suppliers. Although suppliers are a major source of risk on large projects, this paradigm reflects a deeper problem, namely, the belief that business contracts supersede technical quality, a belief that is widespread in large organizations. In short, suppliers may be chosen for qualities other than competence, for example a lower price. This paradigm may be considered part of the conflicting priorities paradigm discussed below.

Another aspect of this paradigm is the belief that developers have no right to expect anything from suppliers other than their product passes a "qualification test." The other view is that the supplier is part of the whole development system, and anything that may affect the performance, operation, and support of the system is a valid expectation.

It is true that the supplier may have proprietary processes and proprietary designs, but the developer can expect to know the "what" of the design if not the "how" of the design.

5.5.5 The Design-Focus Paradigm

This paradigm is the belief that resilience only depends on technical qualities, such as reliability. This paradigm is one of the most engrained paradigms among engineers, who are, by nature and training, technically focused. However, this belief flies in the face of historical evidence. Richard Feynman's (1988) observation that NASA's estimate of the probability of failure was much too low belies this idea.

Nontechnical personnel, such as contracts or supplier management, are not any less vulnerable to this paradigm. They are most likely to view accidents as "technical" problems, not realizing that they, themselves, contributed to them.

This paradigm is at the heart of system resilience. It says that the only important thing is that we have a good design. The management and organizational aspects are not relevant. However, the management and organizational aspects are crucial to the questions: What are the processes for assuring that the "little" problems are uncovered and resolved? What are the processes for assuring communications and quality in the operations and support activities of the system? What is the closed-loop corrective action process? Many other management and organizational processes are critical to system resilience.

This paradigm is particularly detrimental in those domains in which human-intensive systems predominate, such as aircraft, both commercial and military. In these domains humans are significant "components" of the system. This idea is in agreement with the principle that systems consist of "products, people, and processes." In the aviation domain, for example, people perform a significant number of the system functions including piloting and servicing. With this principle in mind, the human elements should be considered in the system resilience equation.

The design-focus paradigm leads to the unsupportable conclusion that system resilience can be quantified in the way that reliability is quantified. Chapter 10, Measuring Resilience, discusses many factors in the quantification of system resilience that are difficult to measure. At best, the numbers assigned to them can only be rough estimates. Leveson (2002) shows that parameters, such as inspection quality, need to be quantified to determine risk level. Hence, the quantification of system resilience can, at best, be a judgmental determination.

5.5.6 The Ethics Paradigm

The ethics paradigm is the belief that if everyone were ethical, there would be no catastrophes. This paradigm has arisen from public commentary on major disasters and is based on the idea that the primary cause of disasters is a failure to maintain ethical standards. Slogans, such as "Do the Right Thing," attest to that fact. Extensive research on these disasters has shown that, although in some cases there were ethical lapses, most disasters were caused by well-intentioned but misguided decisions and other systemic factors, such as poor communications. Vaughn (1996), for example, states that in the Challenger disaster, there was no "intent to do evil."

Citing a NASA study that compares Navy practices with those of NASA, Wald (2003, p. A13) notes that within the Navy "freedom to dissent is a primary element." In contrast, within NASA, engineers who raised concerns were reluctant to express them forcefully enough.

5.5.7 The Distancing Paradigm

One phenomenon often mentioned by researchers, such as Leveson (1995), is that some members of organizations, consciously or unconsciously, avoid any interest in safety issues. The rationale for this is that if a safety issue arises, then they may find themselves the target of a lawsuit or other legal action. This distancing can either be purely personal, or it can be organizational. For example, if the safety organization is separated from the person by many layers of the organizational hierarchy, then distancing becomes easier. The alternative paradigm is, of course, to remain acutely aware of all safety issues and potential safety problems.

5.5.8 The Individual Responsibility Paradigm

Another paradigm often encountered is the opinion that safety is an individual responsibility and that any attempt to focus on systemic safety problems may take the responsibility away from individuals. Although this paradigm may have a certain degree of truth to it, there is no doubt that systemic factors are at work. The alternative paradigm is to realize that if the systemic issues are resolved, then individuals can act correctly as individuals.

5.5.9 The Safety Analysis Paradigm

The next paradigm is one of the most frequently encountered. This paradigm states that human error has already been included in safety analyses. This is an easy paradigm to disprove. Anyone with any knowledge of safety knows that this idea is false. All one has to do is to ask a safety engineer to what extent human error is included in safety analyses. The safety engineer will be the first to admit that it is not possible to account for all human error. On the whole, most safety analyses assume perfect maintenance and perfect operation. Human error may be included in limited predictable circumstances. The alternative paradigm is to recognize that most safety analyses assume well maintained and well operated systems and that issues in the human sectors of the infrastructure should be addressed separately.

Human error in the system is one of the primary contributors to Type B, or internal disruptions, as discussed in Chapter 3, Disruptions.

5.5.10 The Inevitability Paradigm

The next paradigm is probably the most surprising, especially among engineers, physicists and mathematicians, most of whom have studied either reliability or probability in their careers. The idea is that once a probability of failure has been calculated, the system will fail on a certain date. For example, if the mean time between failures (MTBF) is 1000 hours, then the system will fail at 1000 hours. Once that calculation has been made, there is no value in trying to save the system; it will fail.

This is probably the easiest paradigm to disprove. Anyone with a modicum of knowledge about statistics knows that MTBF is a measure of the random nature of an event. An accident may happen before the MTBF value, or it may happen at an immeasurably long time afterward. The fact that this belief persists reflects the general lack of knowledge on this subject even within the technical community.

There are two problems with this paradigm. The first is that MTBF does not predict when something will fail. It only predicts the probability that a system will fail within a certain time boundary. The second problem is that it assumes that this probability cannot be lowered. Even though no one could ever predict a probability of zero, there are many ways to make the probability lower.

5.5.11 The Program Management Paradigm

This paradigm assumes that all issues associated with personnel, management, and organizations should be the purview of program management and that there is no value in trying to connect safety issues (a technical domain) with management issues (a program management domain). Therefore, these issues are outside the purview of system safety. This paradigm is the flip side of the design focus paradigm. It is basically a belief that there should remain a clear distinction between technical and nontechnical factors. When it comes to resilience, there is no dividing line; both technical and nontechnical factors contribute to accidents and both have to be addressed in resilience.

This is just the point of this chapter, namely, that only when we can understand the inter-relationship between the technical and nontechnical aspects in a systemic way will we be able to begin achieving significant improvements in mission success. Almost all textbooks on systems engineering, for example, Blanchard and Fabrycky (2006), devote considerable attention to the management aspects of systems engineering and consider this to be an integral part of the development of a system.

5.5.12 The Risk Denial Paradigm

This is probably one of the most serious of all paradigms. It is simply the refusal of organization members to recognize that there are inherent risks in managing programs. It can also be called the *risk denial* paradigm. The sources of these risks are rarely within the program itself but rather the ambitious schedules, cost goals, and technology objectives that have been adopted. Nevertheless, risks are often, and irrationally, recognized as the fault of the program itself. The alternative paradigm is simply the duty of programs to manage their own risks, not to deny them.

A second reason that the risk denial paradigm is so difficult to manage is that managers will sometimes reply that it is part of their job to take risks, not to manage them. Although most people might recognize a difference between taking risks and managing risks, this difference is not apparent to everyone. This aspect is discussed in the leadership principles at the end of this chapter.

Risk management is one of the most basic of system resilience processes. The existence of the risk denial paradigm illustrates that more is needed besides just a good process; it is imperative that risk management be taken seriously.

The Columbia Accident Investigation Board report (2003) found that cultural factors were at least as important a contributor to the Columbia disaster as technical factors. The NASA culture of accepting risk in the pre-Challenger days was found to be a normative, or established, norm. Hence, risk denial can be said to be a negative cultural factor. Even though most companies and government organizations have risk processes and risk tools, the phenomenon of risk denial is wide-spread as it was found to be in NASA.

It is interesting that despite the Columbia conclusion on cultural factors, the recommendations section (2003, pp. 225–227) contains no recommendations regarding cultural change. All recommendations pertained either to technical or management processes. This fact should not be considered a criticism but rather recognition that cultural change is a subject not well understood or even perceived as achievable.

Vaughn (1996) cites another paradigm, which is a close cousin to the risk denial paradigm. She calls it the *acceptable risk* paradigm. That is to say, over time, the level of risk took on less and less importance. Even though there may have been significant risks, they were just accepted as a standard part of the environment.

One should not overlook the successes of NASA, especially during the Apollo era. It is often asked, what was the secret of NASA's success during that time? There are two possibilities that come to mind: First, the projects were well funded. Second, and most importantly from a culture point of view, there was a significant fear of failure. Feynman (1988, p. 37) has another theory. He points to the atmosphere of "common interest" between the technical personnel and management during the Apollo era which was absent in the Space Shuttle organization.

Risk denial is not restricted to the space domain. Besco (1990 and 1991) discusses risk denial in the commercial aircraft domain. Besco says (p. 2 of Part 1) risk denial manifests itself when accidents are attributed to the irresponsible behavior of a pilot, the other pilots will deny that this type of behavior applies to them. Besco says, "Aviation professionals, including pilots, are so reinforced in the perception that the system is so resistant to risk that a percept of immunity from simple malperformance is wide-spread." Besco says that not one word about risk denial was found in the research for his paper, that is to say, it is not a phenomenon that is widely recognized in the commercial aircraft domain.

5.5.13 The External Constraint Paradigm

This paradigm is the belief that nothing can be done to deal with external factors, such as cost and schedule. Although it may be difficult to deal with these factors, the fact remains that the root causes of disasters often reside in them.

This paradigm is closely associated with the risk denial paradigm. Simply stated, this paradigm insists that there is no way to deal with constraints, such as cost and schedule. They are fixed. Although it may be difficult to loosen these constraints, it is essential to recognize them and deal with them.

5.5.14 The Organizational and Contractual Constraint Paradigm

This is just a single and major constraint of the *external constraint* paradigm. It says that if provisions of the contract with either the customer or with suppliers contribute to a failure to comply with mission success goals, then these constraints cannot be removed or changed.

This paradigm is the belief that one's sole responsibility is to the organization one belongs to and to the contractual boundaries that have been drawn. This paradigm is the belief that contractual constraints cannot also be dealt with. This paradigm is at the heart of the infrastructure system discussed earlier. For a system to be resilient, there should be open communication and cooperation among all the nodes. In adaptability, this attribute is called interelement collaboration.

Another aspect of organizational boundaries is the idea that contracts are sacrosanct. This is probably one of the most difficult paradigms to deal with. But the questions persist: Do the contracts as written constrain the contractor to employ government supplied equipment (GSE) whose quality is unknown or undocumented? Do the contracts exempt the suppliers from responsibility for component defects? These are just two examples. In the end, if resilience is to be addressed, the contracts should also be addressed.

The idea of organizational constraints has entered the popular culture. People who do not have the ability to think beyond their own organization or discipline are said to belong to "stovepipes" or "silos." Although stovepipes and silos are universally disparaged, they nevertheless persist. The goal is to develop a "boundaryless" mind. Ashkenas et al. (1998) had a similar goal in mind in their book *The Boundaryless Organization*. Whereas they were more focused on business objectives, the same principle applies to resilience.

5.5.15 The Random Human Error Paradigm

One of the most common paradigms is the idea that it is not worthwhile to change culture because human nature is unpredictable and unchangeable. Although it is true that there is a large variation in human behavior, it is also true that systems can be defined that are less sensitive to variations in human behavior. The only alternative is to change human behavior, and Senge (1999) has shown that this can be done.

Indeed, the concept of random human error is a primary type of Type B, or internal, disruptions as described in Chapter 3, Disruptions. However, the notion that systems cannot be designed to survive in the face of random human error is ingrained in the minds of a significant sector of the technical community and the public. Hence, the need to define systems that are adaptable to human error, and other random errors, is paramount to resilience, as described in Chapter 8, Resilience Architecting.

It is true that training will reduce human error. However, even the most rigorous training cannot anticipate all the conditions humans have to operate in and decisions that they may have to make. Even then, errors will happen.

5.5.16 The Predictability Paradigm

This paradigm asserts that because there are so very few serious accidents, that is to say, high-consequence events, there is not enough statistical evidence to predict them, and hence efforts to predict them are pointless. Although it is true that there are very few such incidents, most references point to low-consequence events and near misses as indicators of such events. In short, if such low-consequence events can be minimized, then the high-consequence events will also be reduced.

This paradigm has a strong following even within the resilience community. In short, there may be catastrophes that are so random that no indicators would provide a clue. However, as discussed in Chapter 10, Measuring Resilience, the iceberg theory suggests that there may be strong evidence.

5.5.17 The Traditional Paradigm Change Paradigm

Finally, it is argued, people think the way they think and that cannot be changed. So the cultural paradigms having been addressed, the key question is whether the paradigms can be changed. Believers in the traditional paradigm change above would say "no." We will show through the studies of Senge (1999) and others that paradigm change is indeed possible.

5.5.18 The Conflicting Priorities Paradigm

This paradigm is the phenomenon that some aspects of a program take precedence over safety. This is probably the most important paradigm with respect to program management. It basically says that with respect to the believers in this paradigm that when cost and schedule arc constraining, that these factors take precedence over resilience. The results speak for themselves.

This is one of the hardest paradigms to change, namely, if showing a profit is more important than safety, safety will be hard to implement. Leveson (1995) suggests that this paradigm was at work in the Seveso disaster in Italy in 1976 and also the Bhopal disaster in India in 1984. See Chapter 4, Case Histories.

White (2006) relates an historical example of the Conflicting Priorities paradigm. White tells how prior to Eli Janney's invention of the safety coupling in 1868 thousands of railway workers were either killed or maimed attempting to couple freight cars together using the pin system being used at that time. Railway companies were unwilling to change the couplings even at a modest cost until they were forced to do so by legislation. Fortunately, the Janney coupling prevailed and is still being used today.

Flin (2007, chart 11 of 30) defines managerial resilience as:

> Managerial resilience is the ability of managers and supervisors to manage severe pressures and conflicts between safety and the primary production or performance goals of the organization.

Flin (2006) notes several examples of conflicting priorities. First, there was the classic case of the Vasa, the Swedish ship that sank in 1628 on its maiden voyage because the king insisted on proceeding with the ship's launch even though the master shipbuilder told him that the design was unstable.

The second example mentioned by Flin was the Challenger disaster discussed in Chapter 4, Case Histories, in which an engineer was told to "take off his engineering hat and put on his management hat."

Flin also cites Woods (2005, p. 9) regarding the Columbia disaster, also discussed in Chapter 4. Woods states that schedule pressure was strong, "creating strong incentives to downplay schedule disruptions."

Another example is the Metrolink 111 accident discussed in Chapter 4 and also Appendix B. Rohrlich (2008) discusses the conflict between cost and safety in the commuter rail line that collided with a freight train on September 12, 2008 killing 25 people and injuring many more.

Conflicting priorities were evident when, according to Paté-Cornell and Fischbeck (1994), NASA had forbidden the use of risk analysis because it might jeopardize NASA's mission to the moon.

It should not be inferred from this chapter that management harbors more negative paradigms than others. The paradigms identified here are common to all segments of society. They can also be found in many private organizations and government. At the same time, management has greater responsibility to deal with these paradigms. The risk denial and conflicting priorities paradigms are perhaps a greater challenge and greater responsibility to management than to other members of an organization.

5.6 THE GENESIS OF PARADIGMS

Why, it is asked, do we have the paradigms, that is, beliefs that are either obstacles or are demonstrably wrong?

There are basically two sources: First, there is our culture. It is said that our culture is a success-oriented culture. That is our strength. We will attempt things that other cultures may not attempt.

However, our success-oriented culture may also be our vulnerability. If we concentrate on success, is it not possible that we tend to minimize and possibly overlook the weaknesses?

The second source has nothing to do with the cultural environment. It has to do with the work environment. The fact that the work environment is "cost and schedule focused" may work to block out the "problem-focused" environment so essential to mission success.

5.7 SOME APPROACHES

Having identified some major paradigms and their potential negative impact on system resilience, it is now possible to examine some of the more common and forward looking approaches to cultural change. This section will evaluate each approach with respect to the potential effectiveness of each one.

5.7.1 The Training Approach

The training approach is far and away the most popular approach. Training packages are developed and delivered either online or in person. These packages tend to be Microsoft PowerPoint (Microsoft, Redmond, WA) presentations designed to impart the true paradigms to the employees. For the most part, these training sessions do not engage the students. There is rarely any effort to determine whether the students actually understand or agree with the principles imparted. Furthermore, there is little effort to determine whether the principles are actually practiced in the workplace. The sole measure of the effectiveness of the training is the number of students trained. With the above being said, it is safe to say that training as practiced, as such, has had almost no effect on system resilience.

5.7.2 The Charismatic Executive

Another popular notion is the idea that a company executive has the vision and power to change the culture simply by imparting his or her wisdom to the employees. This model is illustrated in Figure 5.2. This model raises some fundamental questions: Is not the executive from the same culture as the employees? What makes him or her free of the paradigms that are held by the employees? In short, there is no evidence that this model has succeeded in improving resilience. Let us examine some other methods that have been suggested.

5.7.3 Socratic Teaching

Socratic teaching is a variation on training. By asking questions, a trained instructor can draw opinions from the class. This method is generally deemed to be better than the conventional training class. A well-trained instructor would be necessary.

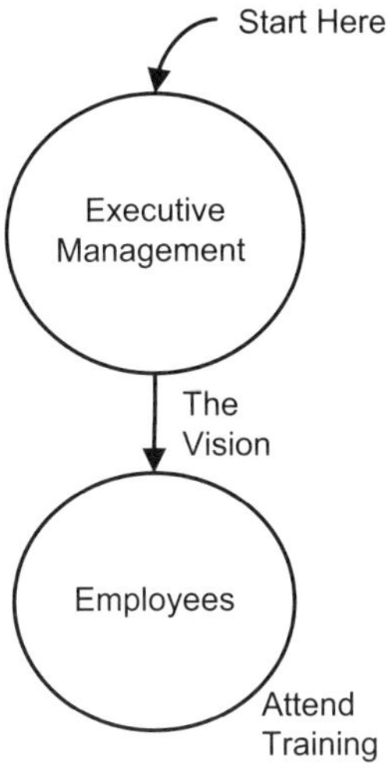

Figure 5.2. The old model.

5.7.4 Teams

Teams are a popular subject in industry and government. Indeed, teams do bring some value. Probably their most important value is breaking down the organizational barriers discussed above under Section 5.5 Common and Alternative Paradigms. If you have a team that consists of a design, production, support, supplier and customer personnel, then communication will greatly improve and organizational barriers will fall. For enterprise systems, teams may be of great value. If there had been a team consisting of all the agencies involved in Hurricane Katrina (see Chapter 4, Case Histories), for example, the cooperation and recovery might have been greatly improved.

From a total infrastructure point of view, as discussed in Chapter 7, Infrastructure a central team consisting of all the nodes of the resilience infrastructure has the potential for being of great value. This team would bring value by coordinating all *the interelement collaboration* (see Chapter 8, Resilience Architecting) among the nodes.

Do teams address all the detrimental paradigms discussed in this chapter? Probably not. They may help in obtaining a common understanding of risks, but erasing the risk denial paradigm is doubtful.

5.7.5 Coaching

Another variation on training is coaching. Coaching is just intensive one-on-one training. The negative aspect of coaching is that it is expensive and can be delivered to a small minority of employees, primarily managers. However, this method is emphasized by Conrow (2000).

Some other suggested methods are not really directed at changing paradigms but rather just making the company run more smoothly and efficiently. A few of examples these are as follows.

5.7.6 Other Methods

5.7.6.1 Independent Reviews. The idea behind independent reviews is to provide an independent and different set of eyes to review all technical work. Typically, reviewers are brought in from other programs. The reviewers are people who have no vested interest in the program at hand. They can interview the engineers and find problems that might not be obvious to those working on the program. Independent reviews are discussed further in Chapter 9, Governance.

5.7.6.2 Cost and Schedule Margins. This one may seem obvious, but the fact is that many, if not most, risks are created by externally generated cost and schedule constraints. And, as was observed earlier, the risk denial paradigm is one of the most common to infect the workplace. If risks are externally removed, that is, by removing the sources of the risk, namely cost and schedule constraints, then the paradigm will be rendered noncontagious.

There are two problems with this approach: First, the persons who set the constraints themselves have to understand the consequences of the constraints. Second, the risk denial paradigm has not gone away. In the end, both of these problems need to be addressed.

5.7.6.3 Standard Processes. This is one of the most common approaches of industry and government. The assumption is that since everyone follows the same processes, then it doesn't matter what their paradigms are; a good system will result. The most famous process model is the *Capability Maturity Model Integration (CMMI)* (2006). Another important process model, especially for system resilience, is the Federal Aviation Administration (FAA) (2004).

In Paté-Cornell (1994, p. 86), B. Buchbinder says that "We are in the midst of a long term effort to change the NASA culture and use PRA [probabilistic risk assessment] to improve programmatic decision-making." Hence, Buchbinder is implying that process improvement will change culture. There may be some validity to this assumption, but it is unlikely that it will change culture. This approach may come under the category of culture management to be discussed below.

5.7.6.4 Rewards and Incentives.

> People do what they are measured to do.
>
> Anonymous

This is also a popular approach to curing the negative paradigm disease. The idea is that employees will be rewarded for doing good things. Weick and Sutcliffe (2001), for example, advocate rewards even for admitting mistakes. The problem with rewards and incentives is the same shortcoming as, for example, the charismatic executive paradigm. It assumes that someone up the management chain understands the good paradigms and how to replace the old paradigms. Although this may be true, there is no guarantee.

In a NASA study comparing their own practices with those of the Navy, the latter being more successful, Wald (2003) notes that NASA had a practice of giving bonuses for on-time launches. According to Wald, this practice raises the possibility that technical quality may have been sacrificed for schedule compliance.

A common practice is to reward executives on the financial performance of the entire organization. One possibility is to provide compensation on the basis of resilience metrics in addition to the normal basis of profit.

5.7.6.5 Management Selection. Although the paradigms listed earlier in this chapter are not exclusively in the domain of management, it is true that managers have a great deal more leverage in the workplace than other employees. Several paradigms have been described that have strong management content: the distancing paradigm, the charismatic executive paradigm, and especially the risk denial paradigm. So what is the answer to this question? One obvious answer is to hire and promote the right managers, that is to say the managers that have the right paradigms.

Once again, there are obvious shortcomings to this approach. As has been pointed out before, before the managers can be measured for their paradigms, someone has to decide what the right paradigms are. In general, these would just be managers at higher levels. Do these managers have the right paradigms? Who knows?

A second shortcoming is that managers are hired for a variety of reasons. There is no evidence that a manager is going to be hired or not hired just because of his or her paradigms.

Finally, is there a standard methodology to test prospective managers for their paradigms? There is none that is well known.

5.7.6.6 Culture Management. One of the more notable attempts to analyze culture and to recommend methods to manage culture is the Baker et al. report (2007) of the 2005 Texas City BP refinery accident discussed in Chapter 4, Case Histories. In this case, the panel concluded that a lax safety culture was one of the root causes of the accident. The panel made four major recommendations, as follows:

- Leadership should take safety more seriously and take steps to improve process safety.

- Provide adequate resources for process safety.
- Incorporate safety into management decision making and provide accountability for safety.
- Incorporate more discipline into processes to improve safety.

Hence, the Baker panel put its entire safety emphasis on leadership. This approach can be called culture management as opposed to culture change. Whether this approach is more effective than the other approaches discussed above remains to be seen. In any event, any approach without management support is unlikely to succeed.

A second example of culture management is the report by the Commercial Aviation Safety Team (CAST) (2007). In short, CAST suggests the following five steps in the management of culture:

- Development of an accident/incident cost analysis tool
- A self-audit process
- Risk management programs
- Revised standard for the Director of Safety (DOS)
- Development of incident reporting and quality assurance

In short, both the Baker committee and the CAST report do not address the prospect of culture change, but rather focus on procedural steps that will help control the effects of a negative cultural environment.

5.8 THE NEW MODEL: SELF-DISCOVERY THROUGH COMMUNITIES OF PRACTICE

Finally, we have come to the point where we have to ask the question: What is the paradigm changing approach endorsed by the organizational psychology community? And second, is there any evidence that it works?

The short answer is that it can generally be called self-discovery and that the way to self-discovery is through so-called communities of practice described by Wegner (1998).

5.8.1 What is the Best Paradigm?

When we look at all the approaches earlier in this chapter, many of them have a flaw in common, namely, that they assume that some wise person or persons knows what the good paradigms are and what the bad paradigms are. Self-discovery does not make that assumption. Self-discovery assumes that each person will find out for himself or herself what the right paradigms are and that the best way to do this is through self-examination in a group setting.

A second benefit of the self-discovery approach is that if a person finds out the right paradigms through this approach, then he or she will internalize the paradigm and truly believe and understand it.

5.8.2 The New Model

Figure 5.3 provides a simplified depiction of the new model, which, as can be observed, is circular. This model should be compared and contrasted with the old model of Figure 5.2. Following is a summary of the similarities and differences.

First, similar to the old model, the action has to start with executive management. The difference is that the executive does not claim to know what the right paradigms are. He or she is willing to find out personally. The executive merely has to endorse the principle of self-discovery and the actions to implement it. Of course, some funding may be necessary to provide facilitators and learning time, but that is the initial investment.

The second step is to establish communities of practice. Some details of communities of practice are given later. Most companies have training departments through which such communities will be organized. However, the term "training" does not officially apply to communities of practice. Most organizational psychologists would prefer the term "learning." According to former Shell Oil CEO Phillip Carroll (1999), "training takes place through repetition and manipulation; people can build computer programs to run seismic records, but not to deal with completely unpredictable circumstances." So the outcome of the communities of practice is learning, not training. According to Reason, "a safe culture is a learning culture."

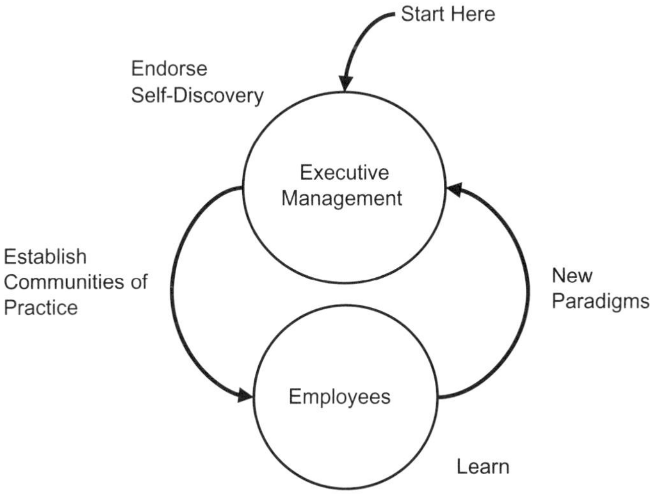

Figure 5.3. The new model.

Finally, the loop comes full circle. It was observed earlier in this section that the executive did not claim to know the right paradigms, so how does he or she absorb the right paradigms? The answer is simple: the same way that everyone else does, through communities of practice.

5.9 COMMUNITIES OF PRACTICE

Now that the concept of self-discovery is understood, it is also necessary to understand the concept of communities of practice, that is the mechanism through which self-discovery takes place.

Although the concept of communities of practice is described in detail in several books, the principles can be outlined as follows.

First of all, communities of practice are a bottom-up concept. Although the executive endorsed the concept of communities of practice, the organization and energy come from the members themselves.

Second, they are informal. There are no PowerPoint presentations (except as the group may find them useful), and a lecturer is not available to provide the truth to the members. The dialog is two way.

Next, the community of practice starts with a core group, perhaps only four or five members. The idea is that the community of practice will grow and continue to bring in more members as the interest grows. Figure 5.4 illustrates the growth of a community of practice from a core group.

Other essential rules include total respect for each other. All ideas are listened to and discussed, perhaps with the aid of a facilitator. The idea is to drive to the core of an issue or a problem and come to a consensus based on logic rather than emotion or past paradigms.

Another rule is that communities of practice should be inclusive. They should cross organizational boundaries and layers. In short, communities of practice provide an environment in which learning and self-discovery can take place.

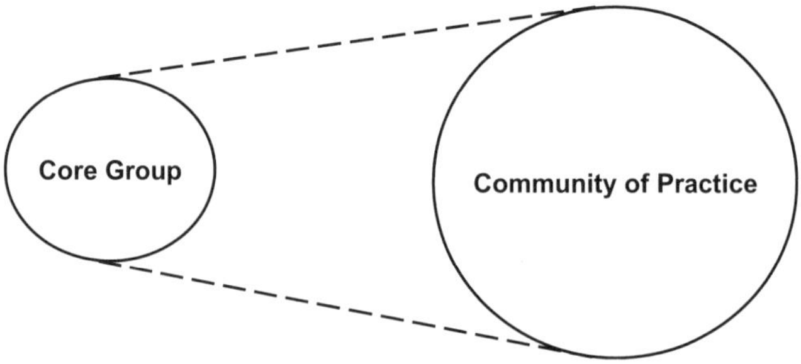

Figure 5.4. Growth of a community of practice.

5.10 CASE STUDIES OF CULTURAL CHANGE

The next logical question is: Has anyone been successful in changing a culture? There are many enterprises that make this claim. The one thing that most of these cases have in common is that the measures of success are primarily from a business point of view, that is, financial. It has not been possible to find any cases in which the cultural change had its focus on system resilience. So, one has to assume that the methodologies that apply to financial success will also apply to system resilience. There is no alternative at the moment; there is simply not enough data to confirm that this extrapolation is valid. However, there have been several doctoral studies that have focused on this methodology for cultural change and that have proven successful.

5.10.1 The Royal Dutch/Shell Case

One of the most notable cases of cultural transformation is with the Royal Dutch/Shell Company. Cor Herkströter, who is a former chairman of the Committee of Managing Directors, led this company through the cultural change process in the late 1990s. Although Herkströter et al. (1999) do not use the term "communities of practice," their description of the methodologies is full of terms like "dialog" and "self-examination." Following this cultural transformation, Royal Dutch/Shell initiated organizational restructuring and other changes to achieve the results it was looking for.

5.10.2 The Xerox Case

According to Senge (1999), the one company that has been successful in implementing communities of practice is Xerox, especially at its Palo Alto Research Center. According to Senge (1999), communities of practice exist throughout all organizations at Xerox. Seely Brown, the director and chief scientist at the research center, says that communities of practice are "the critical building block of a knowledge based company."

5.11 THE PETROSKI LESSONS

Dean (2006) writes about the work and observations of Henry Petroski on the subject of failures and their causes. This chapter summarizes these lessons because so many of them are psychological, that is to say, cultural in nature. They mirror many of the paradigms described above. Following are the lessons:

- Success masks failure. This is basically a restatement of the Titanic Effect described above. Dr. Petroski sees the Tacoma Narrows failure as an example of the paradigm of invulnerability, that is, the Titanic Effect. In addition, Chapter 3, Disruptions, shows that it is an example of a Type A, or external, disruption.

- Systems that require error-free performance are doomed to failure. Challenger and Chernobyl are only two examples of this principle. However, this principle reinforces the Design Focus paradigm, above, namely, that success is merely the domain of technical expertise.
- Computer simulations and other methods of predicting whether components will fail are themselves vulnerable to failure. There is a saying in the engineering world: there are no computer failures, only human failures. What Petroski is saying is that computer simulations are only as good as the humans who create them, which, we have observed, are imperfect.
- Devices can be made foolproof, but not damn-fool-proof. This lesson is a testament to human error and to Type B, or internal disruptions as discussed in Chapter 3, Disruptions. Even a well-designed system may fail in the face of a human error as in the case of the Clapham Junction disaster, in which a highly motivated technician had incorrectly wired a signal box.
- Today's successful design is tomorrow's failure in that expectations for technology are continually on the rise. The demands for technology expectations are shown clearly in the cases of the SpaceShipOne and Goddard commercial space vehicles described above.
- A device designed for one purpose may fail when put to another use.

5.12 OBJECTIVE ASSESSMENT OF CULTURE

The conventional wisdom on culture is that it cannot be measured because it a subjective entity. However, Werner (2001) has developed a model for measuring a safety culture in an organization. The approach is similar to that of CMMI (2006) in which textual criteria are developed in a number of areas. Following are two examples of such criteria with respect to the evaluation of management:

- The [organization] demonstrates a willingness to commit the necessary energy and resources to safety issues.
- The [organization] recognizes and uses the success of previous change efforts and lessons learned.

This approach, when objectively applied, has the promise to provide an insight into the cultural maturity of an organization.

5.13 LEADERSHIP PRINCIPLES FOR RESILIENCE

Almost every organization, government or private, will have adopted a set of leadership principles. These principles appear on posters, in newsletters, and in

Table 5.1. Example Corporate Leadership Principles (The Nestlé Company)

- High commitment to quality products and brands
- Respect for other cultures and traditions
- Personal commitment and courage to take initiatives and risks
- Ability to motivate people and maintain composure
- Curiosity and open-mindedness and commitment to learning and sharing ideas
- Ability to accept and manage change
- Adaptability in thought and deed considering complexity
- Credibility as a result of deed and achievement
- International experience and understanding of cultures
- Flat and flexible organizations
- Clear levels of responsibility and objectives
- Structure with focus on speed and results

training. However, it is safe to say that these principles rarely ever address the question of what an organization has to do to make its product or the organization itself resilient to disruptions.

It is not possible to review all organizations and their published leadership principles, but those of Nestlé (2003) are typical. Like most lists, Nestlé's is a compilation of positive and optimistic principles. These principles are summarized in Table 5.1. This is not to criticize Nestlé; their principles are sound with respect to a business perspective. However, like most organizations, the resilience focus is not high priority.

The following principles are either recapitulations of findings elsewhere in this book, deductions from the general principles of resilience, and the observations by experts in the field.

5.13.1 Place Safety First

According to Flin (2007), one of the top responsibilities of management is to reward "decisions favoring safety." There are several ways that a leader can increase the priority of safety.

First, the leader can get close to safety issues. To stay on top of safety issues, the leader should stay close to them. The leader should consult safety experts and other personnel on a regular basis. As Leveson (1995) points out, if there is a safety issue, resources need to be allocated. That is the responsibility of leadership.

The second thing the leader can do it is to increase the scope of safety. Typically, in product-centered infrastructures, safety is often regarded as a technical issue, whereas organizational aspects are ignored. Reason (1997), for example, devotes a whole book to the organizational causes of accidents. The leader can expand the scope of the safety function, whether it may be a civil infrastructure or a product-centered infrastructure to assure that this function addresses all aspects of safety, both technical and organizational.

Finally, as we saw in Chapter 7, Infrastructure, the organization position of safety, in its expanded role, may influence the effectiveness of the safety function. A simple method of implementing this priority is to make the safety function report directly to the leader. The method is more than just the position of safety on the organization chart; it is the daily contact and interaction of the safety lead with the program leader, and most importantly, it enhances the trust between the program leader and the safety lead.

5.13.2 Focus on Risk

Risk is rarely discussed in management principles. Nestlé, does not mention risk, except in risk taking. It is common for the leadership principles to say that leaders should take risks. Taking risks is only half the story; risks should be recognized and managed. Focusing on risk is more than just following the risk process. The top priority should be to take risk seriously, as discussed in Chapter 5, Culture. Second, the leader should encourage the organization to pursue some of the more advanced aspects of risk discussed in Chapter 6, Capabilities. This pursuit will involve a more detailed look at sources and mitigation of risk. Finally, and most importantly, the leader should encourage examination of what has been called "off-the-table" risks. Organizations tend to ignore these risks because they point a finger at an organization, a customer, or a person. Although these risks may be difficult to treat, ignoring them may have catastrophic results.

5.13.3 Fight Complacency

We saw in Chapter 5, Culture, that it is all too easy to assume that simply because there have not been any recent accidents, there is not one just around the corner. This paradigm was known as the Titanic effect. On the contrary the prudent executive will look for increasing ways to avoid accidents.

5.13.4 Defer to Expertise

It is not the job of leaders to be the experts in every field. However, as pointed out by Weick and Sutcliffe (2001), it is the job of leaders to know who the experts are, to consult them, and to defer to their opinions, especially if there is an issue that has the potential for a risk aspect.

5.13.5 Adopt the Systems View of Your Organization and Associated Organizations

The idea that your organization, your customer, your suppliers, your maintainers and other associated organizations are part of a single system is not an easy idea to comprehend or accept. It will be seen in Chapter 7, Infrastructure, that one of the main challenges of systems of systems is that no one is in charge. This challenge means that some entity should take the initiative to help all the systems within the system of systems work together as a single entity. Your system can adopt the role of the lead system in the system of systems, or at least, it can collaborate with the other systems to create a resilient system of systems. If you are a police chief, for example, you can establish working relations, written or not, with the fire protection organization, the transportation segment, the health segment, power authorities, and the military, as well as the local, state, and national government. This interaction will fulfill the *interelement collaboration* heuristic discussed in Chapter 8, Resilience Architecting.

5.13.6 Encourage Vertical and Horizontal Communication

This principle is the result of the resilience attribute of interelement collaboration discussed in Chapter 8, Resilience Architecting. First, let us examine vertical communication. Most organizations are organized vertically, that is, employees report to managers, managers report to senior managers, senior managers report to executives, and so forth. In this type of organization, communication between the employee and the executive is difficult. Even if the executive announces an "open door" policy, the employee may encounter a negative reaction from the intermediate levels of management.

The executive can do two things: First, he or she can reduce the number of organizational layers. Second, he or she can make it clear to the intervening layers of managers that they will not impede communications.

The barriers to horizontal communications are often called "stovepipes" or "silos." Although there is almost universal condemnation of stovepipes and silos, they have not gone away. Chapter 7, Infrastructure, discusses the importance of communication and collaboration across nodes of an infrastructure, either within an organization or among organizations. Civil infrastructure systems are traditional collections of silos as was observed in the case of Hurricane Katrina, discussed in Chapter 4, Case Histories.

5.13.7 Promote Bottom-up Cost and Schedule Estimates

Cost and schedule risks are most often incurred when the goals are established top down. Top-down goals often ignore the realities and necessities of the bottom-level tasks. The result is that critical bottom-level tasks may be eliminated or reduced.

It cannot be denied that top-level goals may be determined by market forces or governed budget limits. Nevertheless, the risks that may be incurred by these limits should be kept in mind. In the end, project cancellation may be a preferred option to a high-risk option.

However, it should be kept in mind that many resilience aspects bear little or no cost. Good communications come as one low-cost aspect.

5.13.8 Trust Metrics and Early Signs of Trouble

It will be observed in Chapter 10, Measuring Resilience, that there is strong statistical evidence that early signs can warn of a brittle system. Beyond that scientific conclusion, many observers noted that there were often early signs that a disaster was possible.

So what should the leader do? The simple answer is to address the warning signs. Even if the signs are only qualitative, many things can be done. For example, there can be changes in organizational priority, or training may be needed. Even if funding is limited, existing funding priorities may be adjusted. As a general rule, the advice of Weick and Sutcliffe (2001) to pay attention to small problems will be a good first step. See Finger in the Dyke (pp. vi and xi).

Most leadership lists contain an item on ethics. There is no disagreement that an ethical approach is the best approach. However, do other members of your organization really believe that this goal is being pursued? If they do not believe, then accomplishing it will not be likely. The answer to this question can be found through standard survey methodologies. If you find that the trust is low, then corrective action should follow.

5.13.9 Take Responsibility for the Management Aspects of Resilience

We saw in this chapter that a common paradigm among managers is that safety and resilience are technical subjects and not the concern of management. However, we have seen many cases in Chapter 4, Case Histories, in which management decisions, or indecision, played a role in accidents. The idea that an entrepreneur has a vision, and all the engineers have to do is to carry it out, is contrary to the evidence. The entrepreneur, executive, and other management people have just as much responsibility for resilience as the engineers do.

To accomplish this goal, the system view of resilience is necessary. It cannot be assumed that resilience is only the function of technical people, management people, operators, or maintainers. Every person is an element in the resilience system.

5.13.10 Strive for "Commonality of Interest" with Technical Personnel

Feynman (1988) theorizes that "common interest" is the most important aspect of the relationship among managers, engineers, and scientists. His theory is that this aspect is the reason that the Apollo program was a success and that the lack of it was the reason for the Space Shuttle disasters. To this end, leaders should not see themselves as overseers of engineers and scientists but rather as partners in a great endeavor.

5.14 CONCLUSIONS

The paradigms reviewed in this chapter have been ones cited by many references or observed by the authors. Some of them are clearly scientifically and mathematically unsound, for example, the inevitability paradigm. Others just fail the test of basic logic, for example, the risk denial paradigm. Some of them may seem like rational actions to take for the purpose of self-protection if for no other reason, for example, the distancing paradigm.

However, it is not the intent of this chapter to tell you that these paradigms are wrong. If the text seems to imply that they are wrong, then that is just the opinion of the authors, not necessarily the final word. In short, no one can tell you or convince you that they are wrong; you have to discover that yourself.

This chapter has reviewed many different approaches to dealing with these paradigms. Some of them deal directly with mind changing, for example, the self-discovery through communities of practice. Others simply avoid the issue and strive to lessen the impact of destructive paradigms by, for example, providing greater cost and schedule margins.

Two things are clear: The first is that the world has observed an era of unprecedented catastrophic accidents, such as Challenger, Columbia, and Chernobyl. Many respected researchers have concluded that the destructive paradigms were a major contributor to these accidents.

The second thing is that paradigms are hard to change. This chapter has reviewed some of the more common solutions and has focused on one that seems to have the most promise, namely, self-discovery through communities of practice.

Because we are talking about the success or failure of systems, then it seems logical that organizational psychology should be considered an essential discipline within the cross-disciplinary role of system resilience.

5.15 FURTHER EXPLORATION

1. Describe several paradigms not listed in this chapter and how they might be detrimental to the goals of a system.
2. Describe ways in which paradigms can be rendered harmless in an organization.

Chapter **6**

Capabilities

We saw in Figure 2.1 that an organization should have certain capabilities in order, first, to be resilient itself, and second, to create resilient systems if the organization is a product-centered enterprise. Hence, these capabilities apply either to human-intensive systems, such as civil infrastructures, or to organizations that design and produce technological systems. Figure 2.1 breaks these capabilities down into two broad categories, managerial and technical. The premise of this model is that both technical and managerial capabilities are required and that they are required to act as a single function, not two independent functions.

The idea that an organization should possess a basic set of capabilities to develop a system has gained much attention in recent years. The Carnegie-Mellon Institute, for example, has published a model called the Capability Maturity Model Integration CMMI® (2006) that lays out what these capabilities should be. The capabilities in this book, however, concentrate on methods to enhance resilience. CMMI® initially was developed primarily with product-centered enterprises in mind. Although architecture development is discussed, its application to system resilience is not explicitly addressed to the extent it is in this book.

Finally, two earlier papers by Jackson (2002) and Jackson and Hann (2004) derive a set of capabilities directly from the root causes of major accidents as determined by a wide range of research efforts, for example, by Reason (1997), Leveson (1995), Paté-Cornell (1990), Vaughn (1996), and the Columbia Accident Investigation Board (2003). These latter sources provide a set of capabilities that apply across many domains.

The idea of using root causes to establish the required capabilities assumes that all future catastrophes will have the same root causes as the ones in the past. This is not always the case. Leveson (2002), for example, points out that

Architecting Resilient Systems: Accident Avoidance and Survival and Recovery from Disruptions.
By Scott Jackson

approaches should be developed that anticipate future catastrophes even though past catastrophes may result from different root causes from the ones in the past.

The capabilities discussed in this chapter are intended to cover those capabilities not generally found in organizations. Although some of them, for example the analytic methods, may be found in organizations to some extent, the capabilities here are intended to the issues of depth and breadth needed for resilience. Of course, capabilities, such as systems architecting, go directly to the core of defining resilient systems.

6.1 SYSTEM RESILIENCE CAPABILITY DRIVERS

It is logical to ask: What is it that determines the desired resilience capabilities of a system? There are four drivers.

6.1.1 Type of System

As discussed in Chapter 2, System Resilience and Related Concepts, there are various types of systems that may experience a lack of resilience. This lack of resilience is also called brittleness. Therefore, it is logical that each type of system will demand different capabilities. Some capabilities will apply to all systems, and others will be system specific.

For example, all systems require holistic approaches, as described later in this chapter. Human-intensive systems are in special need of these approaches. These include organizational systems and systems of systems. Systems in which humans and technological components interact also demand holistic approaches.

Human-intensive systems have a special need for interelement collaboration and decision-making capabilities.

All systems require managerial capabilities. However, infrastructure systems in which a technological product is involved are in special need of managerial capabilities. Among these managerial capabilities is a cultural initiative capability, which is applicable, of course, only to humans.

Technological systems are most likely to be in need of analytic capabilities, but even these are in need of holistic capabilities.

6.1.2 Type of Disruption

Chapter 3, Disruptions, describes two types of disruptions: Type A disruptions are external disruptions that lead to a loss of function. Type A disruptions include a change in environment, such as a hurricane. Type B disruptions are internal systemic disruptions. They include internally caused loss of function, capability or capacity. Computer and internal human errors are examples. Hence, a safety capability is required to counter both types of disruptions. This section describes more advanced safety approaches to address this aspect.

Large human-intensive systems, such as civil infrastructure systems, are particularly vulnerable to Type A, or external disruptions, such as hurricanes, tsunamis, earthquakes, and tornadoes.

Internally generated anomalous inputs and anomalous outputs can both be categorized as Type B disruptions as described in Chapter 3, Disruptions. Holistic approaches are the primary methods for dealing with these types of disruptions.

6.1.3 Resilience Phases

As described in Chapter 2, System Resilience and Related Concepts, three resilience aspects are as follows: avoidance, survival, and recovery. Each capability will address one or more of these aspects.

Most capabilities will address all three aspects; however, there are a few that only address one or two. Culture and managerial capabilities, for example, address all three. However, holistic approaches and decision making are especially strong in dealing with recovery. Analytic approaches are more applicable to the survival phase.

6.1.4 Resilience Attributes

We saw in Chapter 1, On Resilience, that the primary consideration in the design of a resilient system, either human-intensive or technological, is its adaptability. There are four attributes: capacity, flexibility, tolerance, and interelement collaboration. These attributes are based on the five principles of Woods (2006b). In addition to the considerations of culture and risk, the focus of any organization is to achieve these four attributes.

6.2 DEPTH OF CAPABILITIES

Two issues are on the table here. The one discussed above pertains to which capabilities are of interest and the factors that drive the selection of those capabilities. The other more difficult issue pertains to how much effort and cost should be expended in developing those capabilities. This issue has to do with the depth of the capabilities. Chapter 12, Implementation, discusses this issue at length, but in short here are some guidelines.

At first, it would seem that the probability of a disruption would be an important consideration. However, this consideration is more complex and less intuitive than meets the eye. First, Chapter 3, Disruptions, shows that although an individual disruption may be very small, the probability of the interaction of system elements is higher than generally thought. Chapter 12 also provides three other considerations in the decision to invest money and effort into the development of capabilities. The first is that some capabilities may be implemented at no cost or negligible cost. The second is that some systems are high consequence and high risk so that they deserve more attention. The third is that benefit-cost analyses, such as those conducted by the FAA and discussed in Chapter 11, Cost, can provide guidance as to how much capability is required.

In short, this chapter does not suggest that all capabilities should be implemented on all systems in all scenarios. These capabilities can be shown to be more effective in different scenarios.

6.3 MANAGERIAL CAPABILITIES

Chapter 2, System Resilience and Related Concepts shows that system resilience is dependent on both managerial and technical capabilities, not independently, but rather together as an integrated function. To that end, the following list of capabilities is intended to address the managerial side of system resilience implementation but within an integrated managerial-technical context. As described at the beginning of this chapter, managerial capabilities apply to almost any type of system, any type of disruption, and any aspect of resilience.

6.3.1 Managerial Oversight

Regardless of whether the organizational system is a government agency, a hospital, or the technological product-centered infrastructure, a mechanism is required to assure resilience of the system.

Chapter 7, Infrastructure, describes the system resilience oversight both from the point of view of an organization and from the larger infrastructure. That chapter suggests organizational entities to be responsible for that oversight. This entity would contain all functions that directly affect system resilience, such as system safety. In smaller organizations, this entity might be a single person cognizant of the principles of system resilience.

When the system is a system of systems, as in the case of a large group of organizations, such as local, state, and government agencies, then a mechanism would need to exist to facilitate resilience among all the nodes of that infrastructure. The conventional wisdom is that a collaborative effort would be required. At a minimum, all the nodes of the infrastructure would need some degree of expertise. In addition, a council or committee may be required to coordinate efforts among the nodes. Whether this group is informal or formal, coordination is essential.

6.3.2 Interelement Collaboration

As we saw in Chapter 1, On Resilience, one of the basic attributes of resilience is interelement collaboration. Interelement collaboration is more than communications. It is the ability of two components of a system, human or technological, to share information and resources, if possible, to solve a common problem, that is resilience to a disruption. We have seen that in the case of the New York Power Restoration case, as discussed in Chapter 4, Case Histories, interelement collaboration was strong. In the case of Hurricane Katrina, it was deficient. Interelement collaboration was key to the survival and recovery of Apollo 13. All of these cases are discussed in Chapter 4.

History is replete with examples of major accidents caused by simple errors in communication. For example, Reason (1997) in his study of the Piper Alpha (North Sea Oil Disaster) describes how a communication failure between the night crew and the day crew resulted in the attempt to start a valve that was not completely leak free. No example is more poignant than Kopp's (2003) description of Jésica Santillán, who died when the heart transplant system failed to match her blood type with that of the donor heart when she was having a transplant. These are examples of communications errors at the most detailed level.

Information exchanges can occur at any level of detail, and hence errors can occur at the same levels. At the highest level information should flow among all the nodes of the system resilience infrastructure shown in Figures 2.2 and 2.3 in Chapter 2, System Resilience and Related Concepts.

Many organizations are embracing information management systems. These systems are simply large computer databases that contain all conceivable technical and managerial data needed for the program. All members of the program, whatever node they belong to, can access this database.

Hence, mechanisms are available for *formal* exchanges of information. And, it seems that most major programs, regardless of the domain, are using these mechanisms. However, from a system resilience point of view, is this enough? Although many disasters have resulted from failures in formal information exchanges, many of them occur at a lower, detailed information level. Witness the Piper Alpha and the Jésica Santillán cases also discussed in Chapter 4.

6.3.3 Anonymous Reporting

There is a common belief, as discussed in Chapter 5, Culture, under the *ethics paradigm* that all system resilience problems result from the failure of management to listen to the warnings of lower level personnel. The most often suggested solution is an anonymous reporting system in which anyone can send a message to an independent office which will act on it.

The three problems with this solution are as follows: First, how do you assure that the independent office is indeed independent? Second, even if it were independent, how do you assure that the information will be acted on? Finally, how do you convince the employees that it is independent and that the information will be acted on?

First all, case studies do not support the common beliefs reflected in the *ethics paradigm*. Although some fraction of accidents may have been ethical problems, most accidents are simply the result of innocent, yet flawed, actions or information. The Piper Alpha accident (see Chapter 4, Case Histories) resulted from a miscommunication between a day crew and a night crew. There was no intent to do evil.

All the above having been said, the anonymous reporting system is another channel of information that may do some good in some cases. Organizations that employ it should keep in mind the obstacles listed above when developing it.

6.3.4 Other Communication Lines

The term communication is often restricted to communication between the customer and the developer. However, communication is much broader than that. But with respect to the customer-developer link, the focus here has been on a resilience node to be the primary mechanism for enhancing that. (See Chapter 9, Infrastructure.) Another mechanism is the use of joint databases for information. However, the primary mechanism is the destruction of the psychological walls associated with contracts and the barriers they provide. (See Chapter 5, Culture.) The same links need to exist between all the nodes of the infrastructure as described in Chapter 7, Infrastructure.

6.3.5 Vertical Communication

One barrier to communications in organizations is the rule, sometimes written and sometimes not, that any employee cannot talk to a person more than one level above him or her. A simple way to combat this barrier is simply to issue a policy saying that it is not so. Other solutions are the integrated product teams (IPTs) described in Chapter 7, Infrastructure. Another way is the anonymous reporting system described above in this chapter. Reason (1997, p. 217) says that a safety culture favors "face-to-face communication."

6.3.6 Infrastructure Design

While few sources cite infrastructure as a major factor in system resilience, there are several factors that have an effect: As discussed in Chapter 7, Infrastructure, the infrastructure should facilitate communications and collaboration among the elements. Secondly, the infrastructure should reflect the appropriate authority on matters of system resilience. Finally, contractual and organizational boundaries within the infrastructure should not be an impediment to system resilience. Chapter 7, Infrastructure, elaborates on an approach to meeting these objectives.

But the capability of importance here is the ability to create this infrastructure based on the principles of system resilience. Here is where the boundary between managerial and technical capabilities disappears because the person who creates the infrastructure is the systems architect, as described below. This person creates the infrastructure system based on the heuristics and principles of system resilience.

6.3.7 Governance

One of the largest challenges to system resilience is the identification and elimination of these small errors. Following are a few approaches.

6.3.7.1 Audits. Although audits are unpopular in most domains, when identifying small errors, it is necessary to know what kind of information is passing between what parties, how critical it is, and how accurate it is. A program can establish a database of information threads. This database both establishes what data passes between two parties and also among many parties. Figure 6.1 shows a typical information thread. This figure does not depict all possible paths but is meant to be typical of information flows.

The important aspect of audits is that they cannot neglect any piece of information in the system, no matter how small. They should include conversations between the pilot and first officer in the cockpit of an aircraft. They should include information passed among workers on an assembly line. They should include information passed among nurses in the hallways of hospitals.

6.3.7.2 Training, Education, and Self-Discovery. Although it was emphasized in Chapter 3, Disruptions, that human error is rarely the root cause of an accident, and that the resilience of the entire system is the critical factor, it is also true that the performance and behavior of humans is critical to the success of the system. Two aspects of human performance are discussed below: humans on the "sharp edge," that is, humans whose behavior may cause an immediate failure of the system, and humans in a cultural context.

6.3.7.3 Training for Humans on the Sharp Edge. In many domains, the performance of humans is critical to the resilience of the system. Among these domains are air traffic controllers, health care professionals especially in emergency situations, and the operators of aircraft, nuclear power plants, and space craft. Safety in these systems requires more than simply knowing *what* to do, but also being capable of analyzing, predicting, and making decisions in uncertain and unpredicted situations. Following are two examples of the required human capabilities in these environments.

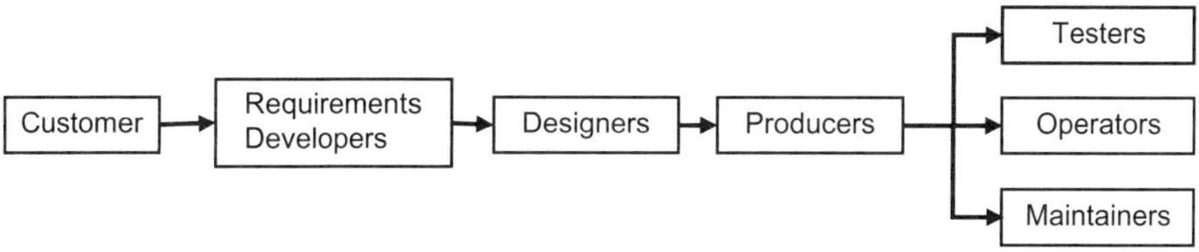

Figure 6.1. Information thread.

Bracco et al. (2008) frame human capabilities necessary in a medical emergency room according to the Skill-Rules-Knowledge (SRK) model. Bracco et al. present these three factors as steps in a hierarchy of complexity in which the practitioner operates. The basic and lowest level is the skill-based level. This level is fast and rigid and involves detailed knowledge of procedures. The rule-based level is more demanding. It is more flexible and resource demanding. The knowledge-based level is slow and time consuming. It requires a lot of cognitive effort and is flexible and creative. Moving from the skill-based level to the rule-based level involves what Bracco et al. call mindfulness. This model calls for the practitioners to move rapidly from level to level. Bracco et al. suggest this model as the basis for training and practice of practitioners. It can also be concluded that this model describes the mental demands on an operator in any sharp-edge domain and can be used for training in that domain.

Another sharp-edge domain is air traffic control (ATC). Like hospitals, ATC systems are human-intensive as defined in Chapter 2, System Resilience and Related Concepts. Malakis and Kontogiannis (2008) analyze the training requirements for ATC personnel in abnormal situations. Cognitive strategies were listed in two groups. The first group was individual cognitive strategies. These included recognition, managing uncertainty, planning, anticipation, and managing workload. The second group was joint cognitive strategies. These included coordination, information exchange, error management, and workload distribution management. Training subjects were tested on a seven-point scale of being able to execute these strategies.

Although the literature has many examples of training methodologies for personnel in sharp-edge domains, the conclusion is the same: Personnel in these domains will encounter complex and unpredicted situations, and strategies need to be developed for how a person should react in these situations.

6.3.7.4 Humans in a Cultural Context. A program can train a person what to say and when to say it. But education is required for the person to understand the importance of information accuracy. Organizational psychologists agree that the most lasting kind of education is self-discovery. A mechanism for facilitating self-discovery is the "community of practice" discussed in Chapter 5, Culture.

6.3.7.5 Independent Reviews. Chapter 9, Governance, discusses the benefits of having third parties review all technical and nontechnical information on a program. This method is intended to counteract the effects of program pressures that may corrupt information.

Independent reviews are a cornerstone capability of system resilience. The key word is *independent*. The premise that independent reviews will catch errors caused either by lack of expertise, lack of knowledge of factors outside the area of expertise of the group being reviewed, and, most importantly, cost and schedule pressures. Hence, independent review teams can identify risks not identified by the design team. Therefore, independent reviews augment and support the risk-management capability discussed in this chapter.

But who are on the independent review team, and what authority do they have? The most robust independent review teams are composed of persons not even within the organization being reviewed. But the most important thing is that independent review team members need to be thoroughly knowledgeable in the principles of system resilience.

With respect to authority, independent review teams can have either advisory authority or thc ultimate authority to assign corrective action items and to approve or disapprove the results. The latter is the most robust case.

When should the independent review team hold a review? Some possible answers are as follows:

- When invited by an organization that recognizes the need
- Before and during any accident investigation meeting

With respect to accident review meetings, an important decision that each accident review board makes is whether an accident is an anomaly (a one of a kind failure) or a systemic failure, that is, when a failure may be a symptom of widespread problems that may be a precursor of future problems throughout the product system, that is the aircraft, spacecraft, nuclear power plant, and so on. Leveson et al. (2006), for example, show that the risk of system failure greatly increases when systemic factors are present. Leveson also shows that the risk of system failure is greater when independent reviews and independent decision making is not present. The value of independent reviews is high.

Wald (2003, p. A13) writes that NASA has conducted a study to determine how it can learn from the Navy, who has a better record in safety. One conclusion was that the Navy has a "robust independent review process," and NASA needs one.

Whatever method is used, information integrity is a top priority for system resilience.

6.3.8 Risk Management

Most sources agree that risk management is a critical capability for system resilience. Leveson et al. (2006), for example, use risk as the primary measure of the effectiveness of corrective actions. The U.S. Air Force instruction (2000) uses the term *operational risk management* (ORM) to emphasize that risk management should cover the entire life cycle of the product. Leveson et al. (2006) also point out that risk management is the control function within the development process in the same way that system components may play the control role within a product system.

In risk terminology, dealing with risks is called risk handling. Risk handling includes incorporating risk reduction steps into the program plan and tracking the risk status. FAA (2004), for example, considers each step to be a separate capability. Although it is not useful to provide a complete description of all the risk management steps, it is useful to identify those aspects that are pertinent to system resilience.

Risk identification generally results in three types of risks: technical, schedule, and cost. Although technical risks are the primary concern of system resilience, schedule and cost pressures, for example, may cause technical shortcuts and thus generate technical risks.

The method of characterizing risk is fairly universal. Risk (R) is defined as the product of the likelihood (L) of a possible event occurring times a quantitative value of that consequence (C), that is

$$R = L \times C$$

The resulting value of R can be classified as low, medium, or high risk. The selected method of handling each risk will depend on the risk level. Figure 6.2 from the International Council on System Engineering (INCOSE) Handbook (2006) illustrates how risk is normally depicted.

This model has an implicit assumption that all risks are independent and linear. Leveson et al. (2006) challenge this assumption. Leveson has done much work in traditional safety analyses, such as fault-tree analyses, and she has concluded that most of these analyses are inadequate because they assume statistically independent risks. On the contrary, she asserts, risks are nonlinear, dependent, and dynamic. Leveson and her team at the Massachusetts Institute of Technology (MIT) have developed a tool called System-Theoretic Accident

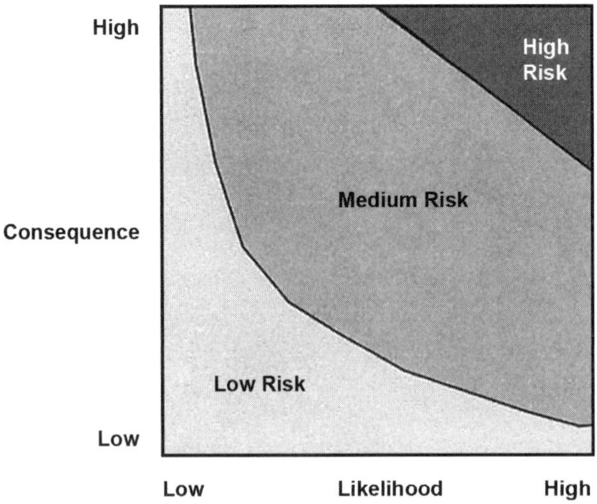

Figure 6.2. Risk diagram (2006).
Source: *Systems engineering handbook: A guide for system life cycle processes and activities* (Version 3, p. 7.13). C. Haskins, (Ed.) Seattle, WA: INCOSE.

Modeling and Processes (STAMP) to simulate these risks. They have created a non-linear risk model that includes technical, managerial and cultural factors. She has shown, for example, that fixing symptoms of problems may lead to much higher levels of risk than fixing the systemic causes of the problems.

As an example, if a broken hydraulic line is discovered to be the cause of a major accident, fixing the hydraulic line itself would not lower the risk much because it is only a symptom. However, if a systemic factor, such as using a low-quality material in all hydraulic lines, was found to be the cause, then fixing this factor would significantly lower the risk.

6.3.8.1 Risk and Culture. However, having a risk process is not the most important aspect of risk. Rather taking risk seriously supersedes all other priorities. The root cause of not taking risk seriously lies in the cultural beliefs, or paradigms, as described in Chapter 5, Culture.

Another view of risk was put forward by Epstein (2006). Epstein noted that in the traditional risk approach all attention is on the upper right hand corner as illustrated in Figure 6.3. He also noted, first, that the solutions to risks in that corner are well known, and second, that most accidents occur because of risks in the lower right hand corner, that is high consequence and low probability. Perrow (1984) makes a related observation, that low-probability, high-consequence failures are the most instructive phenomena for understanding disasters. Perrow also points out that these types of failures are the most useful for identifying methods to prevent, survive, and recover from them. These methods are embodied in the concepts of tight coupling and loose coupling discussed in Chapter 2, System Resilience and Related Concepts.

However, the catastrophe may be caused by a risk that was not even identified. Such was the case in the King's Cross Underground in London, described by Reason (1997), in which a fire in a stairwell killed many people.

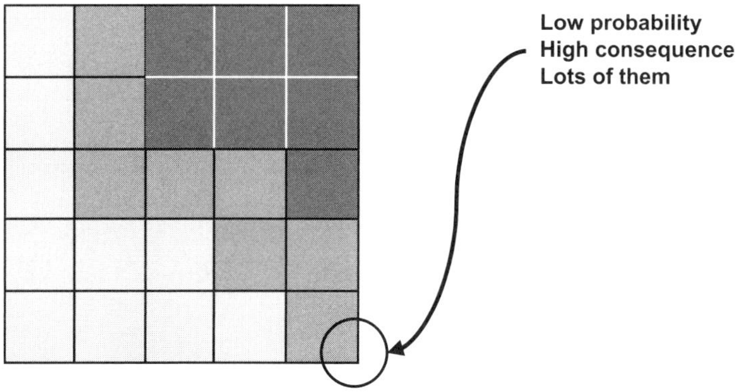

Figure 6.3. Classic risk diagram with Epstein focus.

Another conclusion that can be drawn from the Epstein observation is that the number of low-probability risks must be large to result in the number of accidents that have actually occurred. This conclusion implies that it is the aggregate risk that is important, not individual risks. Vaughn (1996) notes that NASA actually had an aggregate risk process but failed to implement it.

Epstein's observation also points to the law of large numbers discussed in Chapter 3, Disruptions. The law of large numbers, in the context of risk, says that when the number of possibilities of a disruption becomes large, then the likelihood of a disruption will approach a fixed and stable number. When the number becomes large, the composite probability will also become large even if the probability of an individual fault is small.

6.3.9 Cultural Oversight

Many sources discuss the idea that cultural beliefs, or paradigms, play a major role in major accidents. Leveson (1995), for example, discusses the *Titanic effect*, discussed in Chapter 5, Culture. Reason (1997, p. 217) defines the end state of a safety culture. Reason says, for example, that a safety culture "favors face-to-face communication."

Chapter 5, Culture, discusses some of these paradigms and possible methods of dealing with them or changing them. These initiatives are part of the content of the system resilience capabilities. Cultural initiatives are also a governing capability because they influence the quality and robustness of all the other capabilities. Reason's statement about face-to-face communication has a direct effect on the information management capability discussed later in this chapter.

6.3.10 Decision Making

On any program, thousands of decisions are made at all levels of the organization by the developer, customer, operator, maintainer, producer, and suppliers. How can one tell whether these decisions are right and how many wrong ones would be critical? Three general principles of decision making are as follows:

- First, teams should make all major decisions. Chapter 7, Infrastructure, describes the concept of the IPT. One idea behind the IPT is that everyone is represented and everyone has an equal say in decisions. The premise is that if everyone is represented, then the best interest of all parties will be served.
- Second, all important decisions require independent review. Chapter 9, Governance, discusses the concept of independent reviews. The premise behind independent reviews is that when individuals are included who have a view independent of cost and schedule pressures and who have independent technical expertise, better decisions will emerge. Independent reviews apply, for example, to corrective actions (see above), major design reviews, and peer reviews.

- The most difficult, and sometimes most important, decisions are those made by individuals usually in the operational and maintenance phases. The misinformed, but probably well-intentioned, decision by the ship captain in the Texas City 1947 disaster is a good example. (See Chapter 4, Case Histories.)

But how does one make sure that decisions made at that level are the proper ones? There are two answers: one cultural and the other training. La Porte and Consolini as quoted by Reason (1997, p. 214) point to the members of what they call "high reliability organizations (HROs)." They say, "They are driven by a proactive, preventive decision making strategy." But how is such a culture developed? Can it be done by training or by a charismatic leader? Chapter 5, Culture, gives several other approaches including the "communities of practice" approach.

The other approach cited by Reason (1997) is the idea of shifting decision making authority to the lowest level in the organization. Reason cites the policies of the German army as a success story of decision making. Three factors are required to make this happen: First, the first-line supervisors should be selected on the basis of experience, and they require lots of it. Second, there should be a rigid hierarchical structure in which the higher level managers respect the lower level managers. Finally, there should be training, also lots of it.

Maxwell (2009) points out that the San Francisco Fire Department has an escalating decision-making policy. First, as needed, decisions are made at the firefighter level. As the criticality of the incident increases, the decision-making authority rises to higher levels.

One example of decision making, or the lack of decision making, is Hurricane Katrina, discussed in Chapter 4, Case Histories. This is an example in which local agencies were neither trained, empowered, nor supplied with proper resources to deal with the disaster that ensued. This case also shows that reliance on high-level, that is, state and government leadership, proved not to be viable as pointed out by Westrum (2006b).

6.3.11 Corrective Action

Most organizations have corrective action systems to make corrections to problems they encounter. The system resilience teams described above will oversee the corrective action system. Key questions that need to be asked are first, is the problem an anomaly or is it systemic? Second, is it safety critical or not? The cost impact of the answer to these questions can be significant. If a problem is both systemic and safety critical, then extensive modifications may be required. The point is that the questions should be answered honestly and correctly. Otherwise, significant risk will result. The lesson is that independent review of all corrective actions is an absolute necessity.

Wald (2003) notes that the Navy has a "closed loop" corrective action system that requires that all defects be corrected immediately. Wald says that, in contrast, NASA's problem with foam had been recognized for years and that nothing had been done. Hence, NASA would seem to have had a deficient corrective action system at that time.

6.3.12 Schedule Management

Schedule risk is one of the three major categories of risk. Schedule risk occurs when milestone objectives are set that are unrealistically aggressive. This condition occurs when management or customers set unrealistic schedules. This method is called *top-down* scheduling. The risk is often beyond the control of a specific organization, but nevertheless the risk exists.

Schedule pressures are common in enterprises. Marketing personnel and executives often set schedule milestones with little regard to the consequences. They would argue, in their defense, that schedules are determined by "market factors." Be that as it may, these factors are all the more important in the resilience analysis.

The essential ingredients in a robust schedule management capability are first bottom-up assessments of the schedule. For example, the time for the development of a required software module, including coding and testing, may exceed the planned project development period. The risk management capability, above, will determine the steps to handle this risk, including, for example, alternative designs and accepting the schedule slip.

The second ingredient is a critical review of the schedule and the resulting risks. In the most robust case, the review would be completely independent; that is, the review board, in whatever form, would have the authority to approve or reject proposed schedules. In the proposed infrastructure, the program review team would make this decision.

Of course, schedule risks would affect the entire system resilience infrastructure. For that reason, the node representing the customer, developer, maintainers, and all other elements of the infrastructure would address this issue.

Finally, the discussion above focuses on the risks associated with trying to work with too tight a schedule. This subject is central to system resilience. However, the other subject is how to best manage the program once the schedule is established. The most well-known method of doing this is the earned value method. For robust system resilience, this system should be in place as well.

6.3.13 Cost Management

The issues relating to cost management are virtually the same as for schedule management. First, there is the risk of setting development costs too low. This situation happens for a variety of understandable reasons. For example, for commercial products, the market sets the cost goals. For military projects, the government sets the cost goals. All of these pressures result in top-down development cost goals. The consequences of all of these factors appear daily in the public media, namely, reports of excessive overruns. The logic is as follows: cost shortages lead to shortcuts in technical efforts, such as less testing, and so on. Hence, there is a direct effect on system resilience.

The required capabilities are the same as for schedule management, namely, close scrutiny of costs and their effect on technical integrity. The systems engineering integration team (SEIT) and the resilience node would have primary responsibility for this review.

Once again, the management of costs, once the costs are established, would use the earned value system to keep the program on track.

6.3.14 Work Environment

Of all sources, the Federal Aviation Administration (FAA) (2004) has the most comprehensive description of the work environment capability that an organization (a developer, a supplier, a maintainer, or an operator) should have to develop, maintain, or operate a resilient system. In short, the FAA is asking whether the organization has sufficient equipment, procedures, qualified personnel, and other factors. Of course, from a system resilience point of view, the procedures will cover all the capabilities in this chapter. These include the management capabilities, such as cost and schedule management. They also include the technical capabilities, such holistic and analytic methods to be discussed later. The principal work environment areas are work environment needs, standards, and description; qualification of components and personnel; and technology awareness. The FAA document provides details of these areas.

Whereas the FAA document stands alone as a guide for robust work environment management for system resilience, it does reflect real-world realities that might threaten system resilience. For example, is a new company going to have all the work environment capabilities described above? How about a company that is in financial trouble? Is it going to take shortcuts with respect to personnel, equipment and the other factors mentioned above? In short, only a healthy organization with the resources described above can hope to contribute to the development of a resilient system.

The final question pertaining to work environment is as follows: How much margin should be applied to assure that the work gets done and to provide for an adequate margin of safety? As described in Chapter 8, Resilience Architecting, Barter and Morais (1998) address this issue as part of the *margin heuristic* that they say is particularly important in the nuclear industry. It can be said that work margin is a consideration in any high-consequence industry.

6.3.15 Regulatory Environment

Relationships with regulatory agencies can contribute to accidents in various ways. On the one hand, regulatory oversight can simply be absent as it was, according to Hughes (1997), in the Texas City-1947 disaster also discussed in Chapter 4, Case Histories. Or, on the other hand, the regulatory relation could be too "cozy" as quoted by Reason (1997, p. 161), which is also discussed in Chapter 4. In some cases, the operator simply flouted regulations, such as in the case of the Seaview Air Aero Commander in Australia in 1994, according to Reason.

Whatever the root causes of these specific accidents, a system can be resilient with respect to regulatory agencies in three ways:

- The developers and suppliers should develop policies of cooperation with the agencies.
- The regulatory agencies should develop a "detached" relationship with the companies.
- Finally, a mechanism for communication should be established.

6.3.16 Supplier Management

Most enterprises that build large systems, commercial aircraft, spacecraft, and so on, are supplier intensive, that is, suppliers make almost all major subsystems and components. The developer just integrates and assembles them. The quality of the supplier-provided products depends, to a great extent, on the quality of the requirements given to the supplier, the verification done by the supplier, and the nature of the contract with the supplier.

Here you observe the intersection, often collision, of management and technical processes. If price is the primary concern, then technical quality can suffer. If requirements documents to the supplier are not comprehensive, then the product will be deficient. There is a saying in the world of requirements engineering that "If it's not in the specification, it will not be in the system." (See Section 6.4.2, Analytic Methods above in this chapter.) Finally, if suppliers are not required to verify (test, etc.) the products they deliver, there is no assurance that the product will perform as advertised.

The above principles are not new. They reflect the conventional wisdom both in supplier management and systems engineering. However, the field of system resilience brings two new principles, as follows:

Principle Number 1: Suppliers should be an integral and collaborative part of the system resilience infrastructure.

Chapter 7, Infrastructure emphasizes this point. It simply means that suppliers would be an equal partner in all discussions pertaining to the quality of subsystems and components. As Chapter 8 describes, the program review team would be the mechanism for facilitating these discussions.

Principle Number 2: Supplier contracts should not be obstacles to product quality but rather enablers of product quality.

We will see in Chapter 8, Resilience Architecting that one of the heuristics addresses this point directly. The heuristic reads as follows: *The interelement impediment heuristic—There should be no impediments to interelement collaborations.*

Chapter 5, Culture, discusses the contractual constraint paradigm. This paradigm is the belief, held by many, that contracts are walls that cannot be crossed and that if technical quality problems exist, the contract prevents any action on the subject. The fear that changes in contracts will raise costs is rampant. The worst-case scenario is that if the contract needs to be changed to improve quality, then change it and negotiate the cost. If the supplier is not cooperative, then get a new supplier.

However, there are many less harmful alternatives. Using Principle Number 1, above, perhaps the parties can work out a solution within the bounds of the present contract. Second, there may be many solutions outside the contract. For example, most contracts do not address infrastructure subjects, such as communication. Improved communication may improve many situations.

The best-case scenario is to write a good contract in the beginning. From a technical point of view, the contract should reflect the following:

- The supplier should only be put on contract for a comprehensive set of well-documented requirements. The idea that the "supplier will figure out the requirements" opens the door to risk.
- The supplier should be required to conduct all testing and other verification activities, and the results should be signed off by the developer.
- The contract should be worded carefully to assign responsibility in case of operational failures. This is a really difficult thing to do because the operational environment may be different from the environment the product was designed to. Despite all this, the contract should not be viewed as a barrier to fixing the problem.

In short, Principles Number 1 and 2 should assure maximum system resilience from a supplier point of view. The above principles would also apply to contracts between government agencies and developers.

6.4 TECHNICAL CAPABILITIES

Since system resilience covers so many domains, it may be somewhat misleading to use the word *technical.* Technical is not intended to imply that a technological product is involved at all. It is intended to convey the intent of creating and structuring a resilient system, whether that system may be an organization, a technological product or a socioecological system.

6.4.1 Holistic Approach

Nadler and Chandon (2004) distinguish the holistic approach from the reductionist approach. In this book, the latter is called the analytic approach. The holistic approach considers a problem as a whole. The reductionist approach relies on hard data and considers each element of a system separately.

Systems architecting is one of the most critical capabilities in system resilience. Systems architecting is one of the primary methodologies of the holistic approach. Although *systems architecting* may be a new term to many, it has become increasingly well known with the publication of two books by Rechtin (1991 and 2000). In its simplest terms, systems architecting is the discipline that deals with the arrangement of elements within a broader system. In his first book, Rechtin (1991) deals with the architecting of product systems, namely hardware and software. In his second book (2000), he deals with organizations, that is to say human-intensive systems. The importance of systems architecting in system resilience is that it is the primary discipline required for the implementation of adaptability discussed in Chapter 8, Resilience Architecting.

A premise of systems architecting is that many systems are too complex to deal with in quantitative terms. That is to say, the optimum arrangement of elements within the system cannot be determined by quantitative methods. Systems architecting relies rather on heuristics.

Heuristics are not a scientifically verifiable concept; that is to say the systems architect can neither prove nor disprove a heuristic. Nevertheless, systems architecting is widely accepted as the primary way to define the architecture of a system, whether it is an antiterrorism infrastructure or a commercial aircraft. The primary basis of heuristics, and hence systems architecting, is the expertise and experience of the architect. The architect has at his or her fingertips a large number of such heuristics that may or may not apply to resilience. Rechtin provides many. In Chapter 8, Resilience Architecting, Billings (1997) has a few of his own that apply to human-machine interfaces. Billings' heuristics are, of course, based on his own experience. Following are two examples of heuristics, one from Rechtin and one developed specifically with resilience in mind:

The functional redundancy heuristic—There should exist an alternative method to perform each critical function that does not rely on the same physical systems. This heuristic is an adaptability attribute of flexibility in Chapter 8, Resilience Architecting. The best example of this heuristic in action is Apollo 13 in which the crew uses a smaller module for return to earth.

Other methodologies can be considered part of holistic analysis. These include the consideration of culture and risk, for example. Because culture and risk treat the whole system rather than individual components, they can be considered holistic aspects.

Another holistic aspect is the treatment of the unanticipated interaction between system components. Chapter 3, Disruptions, discusses how a disruption may occur even when the individual components perform as designed.

In Chapter 4, Case Histories, the Helios 522, Mars Polar Lander, and Nagoya all occurred because of the interaction between one or more components that worked perfectly. In the cases of Helios 522 and Nagoya, only one component failed; the other component performed as designed. Chapter 8, Resilience Architecting, discusses how these interactions may be treated to make the system more resilient.

Heuristics cannot be considered to be universal. They do not apply to all situations, and they should not be taken literally. They are only a starting point. Some heuristics may seem to conflict; however, when taken together, they can provide some general rules for creating the architecture of a system.

6.4.1.1 The Systems Architect. The final question is as follows: Who is the systems architect? Many organizations now have official positions called systems architect. Other organizations might use terms like chief engineer or program manager. Historically, there are giants like Frank Lloyd Wright. However, it is doubtful that Wright did any architecting with respect to organizations and software, especially because software did not exist. Wright's expertise lay in the design of buildings. In the aerospace industry, names like Kelly Johnson come to mind. Another person who qualifies is John Stevens, architect of the Panama Canal. Stevens had to deal with both the physical architecture of the canal and also disease and morale problems. In any event, these examples do serve to describe the types of architects that are required, that is, persons with breadth of experience and talent.

System resilience places a great deal of demands on the systems architect. In the past, the systems architect was responsible primarily for the physical architecture, such as for an aircraft or a spacecraft. Then as the age of software began to emerge, a software architect also came to the fore. Were the software architect and the physical architect the same person? Sometimes they were and sometimes they were not. Now, with system resilience, a third type of architecture also emerges, namely the organizational architecture.

6.4.2 Analytic Methods

Most writers on resilience, for example Hollnagel et al. (2006) argue correctly that analytic methods alone, within systems engineering, are not sufficient for resilience. These methods, they say, focus on individual components of a system, and many accidents involve the interaction among components that cannot be accounted for in analytic methods. These authors are correct on this point. Dekker and Hollnagel (2006) argue that the Helios 522 accident, discussed in Chapter 4, Case Histories, is an example of such an accident.

However, it can be concluded that for a system to be resilient, it must be dependable as well. And dependability is the subject of analytic methods. Therefore, such subjects as requirements and verification are a necessary prerequisite for resilience, but are not totally sufficient.

Engineering is traditionally regarded as an analytic discipline. That is, engineering processes apply the principles of physics to the design of a technological system. The premise is that a system (aircraft, nuclear power plant, etc.) can be decomposed into tiers of a physical architecture. A basic principle is that requirements are the basis for design.

Of course, this process only applies to product systems, that is, systems that consist of hardware and software. Many systems that have failed are indeed product systems, for example, Chernobyl; the root cause of many of these disasters is a failure to define and implement requirements correctly.

6.4.2.1 Humans and Requirements. So, how does the human fit into the whole requirements scene with respect to catastrophes? Two important answers to this question are described below.

First, it is impossible to lay requirements on humans. This statement is not talking about requirements like: All humans in the system shall be at least 6 feet tall. Yes, this is a valid requirement and can be verified. The requirements in question are those like: The human shall make the correct decision 90% of the time. This requirement cannot be verified for two reasons: First, the conditions under which the decision has to be made are unpredictable. Second, a precise definition of "correct decision" cannot be made. Hence, the best the system architect can do is to use "heuristics" to create a human system, as discussed later in this chapter.

The second and more important aspect of humans and requirements from a resilience point of view is that it is humans who create, implement, and verify the requirements. It is the imperfection in this system that is the root cause of many disasters.

A basic attribute of requirements is achievability as reflected in the technology readiness levels of Table 6.1. With regard to achievability, this quality is the link between requirements and risk.

The shortcoming of the standard requirements process is that it does not account for the unpredicted disruptions discussed in Chapter 3, Disruptions. For this reason, adaptable designs are paramount, as discussed in Chapter 8, Resilience Architecting. Adaptability requires methods beyond the traditional requirements approach. These methods were discussed in Section 6.4.1, Holistic Approach, above.

Chapter 3 also discusses disruptions that could have been anticipated if the analysis had been thorough enough. A number of accidents are of this type. With respect to requirements and resilience, the following principles apply:

6.4.2.2 Requirements in Depth. To handle disruptions, requirements have to be developed to an unprecedented level of detail rarely known. The case of Mars Polar Lander is a good example. In this case, which is discussed in Chapter 4, Case Histories, if requirements had been developed to the strut level and software level on that vehicle, the likelihood of mission failure would have been considerably reduced.

Table 6.1. Technology Readiness Levels

1. Basic principles observed and reported
2. Technology concept and/or application formulated
3. Analytical and experimental critical function and/or characteristic proof of concept
4. Component and/or breadboard validation in laboratory environment
5. Component and/or breadboard validation in relevant environment
6. System/subsystem model or prototype demonstration in a relevant environment
7. System prototype demonstration in an operational environment
8. Actual system completed and qualified through test and demonstration
9. Actual system proven through successful mission operations

6.4.2.3 Requirements in Breadth. Similarly, requirements need to be developed across all "enabling" elements of the system, for example, production and support. This practice is also rare in industry. The example of the ValuJet accident illustrates well this principle. In that accident, which was also discussed in Chapter 4, it is highly unlikely that requirements were propagated across the two layers of organizational boundaries between the developer and the maintenance organization. If it had, the result might have been very different.

Verification is the determination that a product meets its documented requirements. Four accepted categories of verification are test, demonstration, analysis, and inspection. In recent years the term *verification* has come to replace the older term *test and evaluation* because requirements can be verified by methods other than test.

From a resilience point of view, the same principle applies as for requirements above, namely, that requirements need to be verified to a level not common in industry today. For example, if the Mars Polar Lander, which is discussed in Chapter 4, Case Histories, had its requirements verified to the strut and software level, that mission might have been a success. Similarly, if the interface verification of the Mars Climate Orbiter had been verified then the results might have been different.

Of course, verification may require capabilities beyond the ability of the organizations to fund. Most high-quality verification methods do not come free. For example, for verification by analysis, a finite-element analysis (FEA) or a computational fluid dynamics (CFD) simulation would be very expensive. For any aircraft, the cost of flight testing is high. Hence, these factors should be included in the capabilities list.

Kaslow (2004), for example, presents a catalog of factors that contribute to space systems failures. Among these is a list of types of testing that may have been inadequate in failed missions.

Another question that needs to be answered is: Can systems with *unpredicted disruptions*, as discussed in Chapter 3, Disruptions, be verified? The obvious answer is that it cannot because, by definition, the requirements for unpredictable disruptions cannot be identified. The answer lies in the concept of *validation*, that is, to determine whether the system meets the intent of the original mission. This definition implies that validation tests for *missing* requirements, which is actually what it does. This principle explains why aircraft are lost during flight test. This is unfortunate, but it is hoped that validation testing will not end in that way.

As discussed in Chapter 3, Disruptions, the lack of software verification was recognized to be the root cause of the loss of navigation capability on a squadron of F-22 Raptors that had not been programmed or tested for crossing the International Date Line. This is a clear case of a predictable disruption, and hence the error was entirely avoidable.

In a lecture full of such vivid examples, Augustine (1996) describes a failure of a Doppler radar in which the software code was lacking a hyphen in one line of code. This is an example of the importance of testing.

Finally, the discussion above only applies to *product* systems, such as aircraft and chemical and nuclear power plants. Human systems just cannot be verified to a set of requirements the way product systems can. For example, can a hospital system be verified? Of course, there are many products in the hospital system, such as the power system and medical devices. These can all be verified. However, hospitals are human-intensive systems. The only hope for human-intensive systems is systems architecting, that is, the design of the system by heuristics, as discussed below in Section 6.4.1, Holistic Approach. For resilience, the systems architect needs to design the system with adaptability in mind, as discussed in Chapter 8, Resilience Architecting. If all of this is done, then a resilient system will result.

6.4.2.4 Interface Management. An example of the failure to establish an interface management process is the Mars Climate Orbiter described in Chapter 4, Case Histories. Although the failure to use the metric system has received much attention, a deeper look points to the lack of an interface management system.

However, the study of the architecting of resilient systems increases the scope of interface management. Chapter 3, Disruptions, discusses the phenomenon of disruptions caused by the unanticipated interaction between components. Chapter 3 also provides an expanded view of the traditional tool for analyzing interfaces, the N2 diagram. Figure 3.3 shows that the interaction between system components can either be desirable (D) or undesirable (U). In addition, Chapter 3 makes the point that undesirable interactions may occur even if one or both components operate as designed. Chapter 8, Resilience Architecting, discusses various methods for treating these unanticipated interactions depending on the type of system, whether it is a technological or a human-intensive system.

6.4.2.5 Safety. One might ask, "Isn't system resilience all about safety?" Well yes, but not quite. Traditionally, system safety has asked the question: Has the system, that is the product system, been designed to be safe? This is a much narrower question than system resilience addresses. Nevertheless, system safety plays a major role in system resilience. The Presidential Commission for the Challenger disaster, as described in Chapter 4, Case Histories, cites the lack of a rigorous system safety process as a major problem in the Challenger disaster.

For system resilience, however, the safety organization needs to take on an expanded and authoritative role. First, as suggested in Chapter 7, Infrastructure, the safety organization needs to be placed in close proximity to the program manager both physically and organizationally. This proximity gives safety a much greater influence in matters of system resilience. Second, it gives safety the authority to act on matters of system resilience. Finally, a key function of the safety organization is periodic reporting on safety issues pertinent to the program.

A second expansion of roles is to give the safety organization cognizance over matters beyond the design of the product system. The safety organization should deal with organizational issues as well, that is, organizational threats to system resilience. In today's engineering world, this expanded scope is rarely implemented. As Woods (2006, p. 320) states:

> Safety organizations must be involved in enough everyday organizational activities to have a finger on the pulse of the organization and to be seen as a constructive participant in the organization's activities and decisions that affect the balance across safety and production goals.

Finally, Leveson (2002) stresses that a key attribute of a safety organization is its independence; that is, it needs to be able to make safety assessments without any organizational hindrance and to carry those assessments to the highest levels of management.

In addition, Lintern (2003, p. 722) criticizes today's safety approach as "locked into a wrong-headed approach of retrospective analysis followed by development of more intricate control." Lintern advocates, rather, an approach that identifies the global constraints and local requirements and helps designers develop a more comprehensive and internally consistent set of rules that adapt to the situation at hand.

Dekker and Hollnagel (2006) and also Leveson (2002) point to the limitation in traditional safety analysis that is focused on individual components rather than a combination of components, as was the case for the Helios 522 accident discussed in Chapter 4, Case Histories. Leveson also states that traditional safety analysis does not treat the interaction of components, each one of which performs as designed. To deal with these conditions, the attributes of resilience discussed in Chapter 8, Resilience Architecting, should be brought to bear. In particular, the discussion of the *avoid hidden interactions* heuristic is pertinent to the issue of individual components that perform as designed but create a disruption because of their interactions.

6.4.2.5.1 Safety in human-intensive systems. There is no human-intensive system that is more sensitive to safety issues than a hospital. Murphy (2006, p. D10) says that about 98,000 patients die each year from preventable medical errors. This number does not include nonfatal errors, such as amputations of the wrong limbs. The challenge is that the hospital cultural atmosphere is a highly dynamic environment of various people, unlike aircraft. Mistakes are generally categorized as human error, "particularly failures in communication, leadership and decision-making." The similarities between these root causes and those of large product systems are discussed in Chapter 4, Case Histories. Recent improvements have been made in hospital safety through the use of training and checklists similar to those used in aviation, Murphy says.

6.4.2.6 Technology Management. Technology management is the third apex of the risk triangle. With each new advance in technology, comes a risk. Increased demands on performance drive technology, which drives risk. Certainly, going to Mars requires an increase in technology beyond that required to go to the moon. The progression is unending.

But in the end, the capabilities required to manage technology risk are basically the same as those required to manage costs and schedules. It has been said that you can have two of the following three: low cost, high performance, and fast development, but not all three of these.

Like cost and schedule, technology risk occurs when performance goals exceed performance capabilities. Also, like cost and schedule risks, technology risks require close scrutiny by risk teams and by the independent review teams (see 6.3.3.5 Independent Reviews, above).

One tool for evaluating technology is the National Aeronautics and Space Administration (NASA) Technology Readiness Level (TRL) scale shown in Table 6.1, Technology Readiness Levels, also used by the U.S. Department of Defense (DoD) (2001). The idea is simple: If a technology is unproved, then a low TRL is assigned. If the maturity level is high, then a TRL level of 9 is assigned. For example, the internal combustion engine would receive a TRL level of 9. The U.S. General Accounting Office has declared that no system should go into production unless the TRL is at least a level of 6.

In the end, each program needs to evaluate the technology of its systems and determine whether they pose a high risk to development. (See Section 6.3.8, Risk Management.) If so, system resilience will suffer.

To the extent that cost and schedule risks affect technical quality, then these factors are of importance also.

6.4.2.7 Expertise. In the days before complex systems, engineering, and process management, the development of large systems depended a great deal on expertise. Today, processes have become more important. That is to say, if a program has all the processes in place and these processes are executed, then a more resilient system will result. However, as technologies progress, expertise is as important as it ever was.

A lack of expertise can occur for a variety of reasons. First, there is simply the market. If engineers and other experts are going to other companies or to other fields, then expertise will suffer. Next, there is the aging process: As experts begin to retire, they are not always replaced by knowledgeable younger people. Older people bring with them something that younger people do not acquire overnight, that is, experience.

Expertise is necessary both for performing complex jobs and also for reviewing and checking the work of others. Expertise is a necessity for the independent review teams and peer reviews discussed in Chapter 9, Governance. Furthermore, expertise is particularly important for the detailed cross checks also discussed in Chapter 9. A basic ground rule is that every calculation and other operation should be checked "by a person of equal or greater expertise."

So, how does a program retain expertise? There are two answers: First, expertise should be a top management priority. Second, organizations with the long view have generous educational compensation programs.

Finally, let us examine a couple of case studies for which the lack of expertise was a root cause of resilience deficiency. First, there was the Tacoma Narrows bridge disaster discussed in Chapter 4, Case Histories. As discussed in Chapter 4, the root cause of this disaster was that the engineers on the project were unaware of the aerodynamic effects that resulted in the bridge collapse. Although one might argue that it is unreasonable for a bridge project to engage the services of an aerodynamicist with knowledge of obscure effects, such as unstable turbulent boundary layers, the fact is that this disaster did occur for this reason. This case study falls into the risk chart corner described in Figure 4.2, above, that is to say, it had a high consequence and a low probability of occurrence. This case can also be described as an example of a Type A disruption, that is, a disruption from an external source. In short, resilience calls for examining just such cases.

A second case is the Comet aircraft disasters, which was discussed in Chapter 4, Case Histories. Once again, the root cause of this disaster was metal fatigue, this phenomenon is well known to engineers. So why was metal fatigue not known to the designers? The simple answer is that it was considered such a remote possibility that it was not even thought about. This is another lower-right-hand-corner-of-the-risk-diagram case. Or perhaps it was a matter of expertise. Could seasoned structural engineers have been brought to the project to review the possibility of such remote phenomena? The answer is more than possibly: They should have been consulted. The discussion in Chapter 4, Case Histories also reveals that the fatigue problem was exacerbated by the fact that the stress in the structural members was increased by the depressed part numbers on the aircraft parts. It could be argued that even this aspect demands a higher level of expertise.

More recently, Wald (2003) citing a NASA study, which is discussed above, compares the practices of the Navy with those of NASA, where the Navy has a better track record. The study noted that the low launch rate of the shuttle program had a negative effect on the technicians, engineers, and others to retain proficiency. The Navy's large number of reactors, about 100, had a positive effect on proficiency.

6.4.2.8 Software. Although software is a product and not a capability as such, it deserves primary focus within system resilience because it is the genesis of an explosion of complexity in modern systems. Software is the brains in most modern systems, and hence it is the source of much risk. Software is one technology to be the focus of Section 6.4.2.6 Technology Management section above in this chapter. Following are some truths about software that give it this preeminent position:

- Software's complexity makes its development more challenging than other components.
- Software's complexity makes getting it right the first time a rarity. Multiple builds are common.
- Software testing generally requires more time than other components.
- Software is generally a Boolean component, that is, it either works or does not work. There is rarely an "almost" condition.
- Total development time for software usually is the schedule driver for entire programs.

All the above add up to the conclusion that software is risk prone. One source of risk is that software's lengthy development schedule often creates schedule risk for the program (see 6.3.8 Schedule Management above).

Hence, rigorous software processes are essential for system resilience. These processes include, first, a realistic assessment of the software schedule and its effect on the entire program. Part of this schedule is to assure that software receives time for an adequate number of builds and also for adequate testing. Finally, the program needs to assess the true risk level associated with software and to take appropriate steps to handle these.

The critical issue with respect to software is the number of scenarios that have to be tested for any code. Failure to test every scenario results in the risk that the software will fail in an operational scenario that has not been tested. Jackson (2006) describes a set of tools that have been developed to reduce these scenarios to a tractable number. These tools generally employ a SAT (for satisfiability) solver. The SAT solver models the behavior of the system, describes objects within the design, collects like objects, and mathematically links them. In this way, only logically connected objects are tested resulting in a considerably reduced number of test cases. Hence, methods like these promise to improve system resilience.

Software was a major contributor to the Nagoya and Mars Polar Lander accidents discussed in Chapter 4, Case Histories, and also in Chapter 3, Disruptions. Hence, software quality is critical to resilience.

6.4.2.9 Manufacturing. Many capabilities are in place today to help assure high quality in manufacturing. These include manufacturing standards, the use of metrics, change control, and others. Although most of the processes are widespread, the onus on every node of the system resilience infrastructure is to assure that rigor is applied to each one.

But what special degree of rigor does resilience require? The answer lies in the discussions of requirements and verification above. That is to say, in traditional engineering practices the over-the-wall method was used. Resilience requires that manufacturing is an integral part of the system. While this principle is logical, it is difficult to implement in practice. As a matter of fact, it rarely is. But both ISO/IEC 15288 (2002) and ANSI/EIA-632 (1999) identify production as an enabling element of the system. These documents show that the elements should be developed with the same rigor as the "end product," that is, the aircraft or power plant. Hence, the "requirements breadth" principle discussed above should be employed in the manufacturing world.

6.4.2.10 Operations. The conventional wisdom is that when an accident occurs as a result of "human error," particularly on the part of the pilot or some other operator, nothing was really wrong with the system; it was just a misuse of the system. Reason (1997, p. 226) states that "[human] errors are the symptoms that reveal the presence of latent conditions within the system at large." It is the systems engineering philosophy and the system resilience philosophy that the operators are components of the system and need to be treated with the same rigor as the actual hardware and software.

Part of the problem is organizational boundaries. For commercial aircraft, the pilots work for the airline companies. For military aircraft, they work for the military organization. In both cases, the customer is responsible for pilot selection and training. Pilot selection may include such diverse factors as psychological testing and knowledge of English; fluency in English is an international standard for commercial pilots of all countries. A lack of fluency in English has been a contributing factor in aircraft disasters.

Chapter 4, Case Histories, shows that many other types of operational errors contributed to or were the direct cause of many major accidents. These include the 1947 Texas City disaster and Chernobyl. In both the Texas City and Chernobyl cases, errors in judgment by the local authorities played a major part in these accidents. Hence, system resilience to operational errors is dependent on both organizational level integration and on detailed capabilities, such as decision making (see System Resilience Oversight section in this chapter).

In short, the philosophy of the resilience infrastructure as described in Chapter 7, Infrastructure, should become part of the enterprise mindset and implemented with that philosophy in mind. Chapter 12, Implementation, addresses this principle.

Finally, the organizational and contractual constraint paradigm, which is discussed in Chapter 5, Culture, should be addressed. The operational phase of a program should be considered integral to the whole program.

6.4.2.11 Technical Management. Technical management is inextricably interwoven with the technical effort. Many other capabilities listed in this chapter are technical management capabilities.

Technical management is responsible for all engineering activities, such as requirements management, verification, risk management, configuration management, and so on, all of which are discussed in this chapter. The INCOSE Handbook (2006) is a source for these processes.

Technical management has primary responsibility for the cost management, schedule management, and technology management capabilities discussed earlier in this chapter. The failure to manage these aspects will result in major risks.

Technical management also has responsibility for risk management, which is also discussed earlier. Risk management means taking risks seriously and owning the risks. Technical managers cannot afford to be the victim of the risk denial paradigm discussed in Chapter 5, Culture.

Finally, technical management should own the work environment capability, which is also discussed earlier in this chapter. This entails ownership of all processes, equipment, and the means to assure a qualified work place.

From a resilience perspective, a lot of responsibility falls on the technical manager's shoulders. Technical management is the glue that holds all other system resilience capabilities together.

6.4.2.12 Reliability. Contrary to conventional wisdom, major accidents are not caused by a reliability failure, that is the failure of a component to operate simply because it had reached the end of its operational design cycle. In the context of the Swiss cheese model discussed in Chapter 3, Disruptions, reliability can be recognized as just one layer of defense, that is, one layer of Swiss cheese. This is not to say that reliability failures do not occur; they are frequently the cause of a Type B, or systemic, disruption as in the case of Apollo 13. As we discussed in Chapter 4, Case Histories, Apollo 13 survived its reliability disruption because of the resilience of the whole system.

All of this is not to say that reliability is not important. It is a given that all system components should be reliable. For this reason, reliability is one of the basic capabilities that a system should have and that an organization should design for. In Chapter 1, On Resilience, we discussed that resilience occurs in three phases. Phase 2 is the survival phase. An unreliable system will have a difficult time surviving until it can recover in Phase 3. So, yes, reliability is the first line against disruptions, but it is not the whole story.

Leveson (2008), for example, points out that highly reliable components are not necessarily safe and that many accidents occur without component failure. Leveson (2002) points to failures caused by the unanticipated interaction among components that performed as designed. Chapter 8, Resilience Architecting, provides several ways to analyze and deal with these types of failures.

6.4.2.13 Maintenance. According to Reason (1997, p. 85), "There have been a large number of organizational accidents in which maintenance failures were either a significant cause or an exacerbating feature." Reason lists the following accidents as examples: Apollo 13, Flixborough, Three Mile Island, American Flight 191–Chicago O'Hare, Bhopal, Japan Air Lines–Flight JL 123, Mount Osutaka, Piper Alpha, Clapham Junction, and Phillips 66 Company.

Paté-Cornell and Fischbeck (1994) underscore the importance of maintenance by applying the probabilistic risk assessment (PRA) methodology to the Space Shuttle and concluding that a change in maintenance strategy could improve the probability of failure by about 70%. In the same paper, Paté-Cornell and Fischbeck note that previously NASA had forbidden the use of risk analysis because NASA had feared that the small probabilities of success would jeopardize NASA's mission to the moon.

Traditionally, maintenance is an afterthought in the design process. The thought process is as follows: "now we have designed a system, how do we maintain it?" Maintenance requirements are documented in a *technical order* and sent to the manual development team. On the whole, the process has been of the "over the transom" type.

Modern thinking, especially with respect to system resilience, is that the maintenance system is an integral part of a larger system that includes the end product (the aircraft, space craft, or nuclear power plant), the support system (to include maintenance) and six other enabling systems according to ISO 15288 (2002). Hence, the maintenance system should be designed and operated integrally with the rest of the system.

One impediment to a robust maintenance system is organizational boundaries. Systems are either maintained by the customer, a supplier to the customer, or a separate department within the customer organization. Rarely does the developer maintain the system it develops. Organizational boundaries present barriers to close cooperation and collaboration between the developer, the operator and the maintenance organization.

The keys to a robust maintenance system are first, a robust communication system (See Section 6.3.2.2, Other Communications Lines) and, second, an infrastructure that treats the development, operations, and maintenance systems as elements of a single integrated architecture (see Chapter 7, Infrastructure). These factors will enable the resilience architects to assure that maintenance requirements and approaches are treated together and not at arm's length.

Most importantly, the *requirements breadth* principle should be employed as described above in Section 6.4.2.3, Requirements in Breadth. As for the production element described above, the support element is also an integral part of the system and demands as rigorous a requirements and verification process as the rest of the system. Unfortunately, this is rarely the case, and resilience suffers.

6.5 FURTHER EXPLORATION

1. Provide a summary of the expanded scope of each item in this chapter that would be demanded by system resilience. These include: oversight, risk management, supplier management, systems architecting, maintenance, regulatory environment, technical management, information management, work environment, operations, manufacturing, software, expertise, system safety, requirements, verification, technology management, cost management, schedule management, and cultural initiatives.
2. Using case studies from Chapter 4, Case Histories, or from your own research, explain how any or all of the above capabilities might have played a role in increasing system resilience.

Chapter 7

Infrastructure

In the context of resilience, the word *infrastructure* can have several meanings. In some cases, the infrastructure system is the system itself that is the system whose resilience should be maintained. As discussed in Chapter 4, Case Histories, Hurricane Katrina is an example in which the infrastructure was critical. The New York Power Restoration case is another example discussed in Chapter 4. In both these cases, the infrastructure is not just a single organization; it may be many organizations. Of course, in both cases, there were hardware elements, such as generators, bridges, and so on. But the system that mattered was the organizational infrastructure. In the context of the critical infrastructure protection (CIP) program (2007), the infrastructures of interest are the infrastructures that are vital to the economic well being of a country. These include, for example, telecommunications, transportation, power, water, fire protection, and the medical infrastructure.

As discussed in Chapter 2, System Resilience and Related Concepts, and described by Pommerening (2007), when these types of multiorganizational systems become resilient, they can be called complex adaptive systems (CASs).

The second major category is the product-centered infrastructure. In this case, the infrastructure system is the combination of developer, customer (military or commercial), operators, and maintenance organizations. However, it can be said that if the infrastructure is brittle, then the product is also likely to be brittle. Hence, it is valuable to study both the resilience of the product itself and the infrastructure that produces the product.

7.1 THE ARCHITECTURE OF AN INFRASTRUCTURE SYSTEM

According to the Department of Defense Architectural Framework (DODAF) (2007), any system can be defined by its *architecture*. Although the DODAF

Architecting Resilient Systems: Accident Avoidance and Survival and Recovery from Disruptions.
By Scott Jackson

was developed by a military entity, the principles can be applied to any domain or type of system. Each architecture has a *view*. The following paragraphs below discuss two infrastructure views: the operational view and the organizational view.

7.1.1 Operational View

The operational view is primarily an exterior view. That is, this view depicts the relations between an organization and other organizations and entities. This view is particularly important in depicting systems of systems, such as a civil infrastructure in which the relation between, for example, the power node and the telecommunications node is important.

One of the more useful views, especially from a system resilience point of view, is Operational View Number 2 (OV-2) from the DODAF. Figure 2.2 shows a typical operational view of an infrastructures system focused on achieving system resilience. This figure shows an infrastructure in which the infrastructure system is the system of interest. Hurricane Katrina and the New York Power Restoration cases are examples. The "nodes" of this system are the various entities that comprise the infrastructure. The nodes include, for example, the power node, the transportation node, the telecommunication node, the law enforcement node, the life support node, the government node, and the human node, that is, the public. Other nodes are possible, such as the fire fighting node.

Appendix A, Domain-Specific Example for Architecting a Fire Protection Infrastructure provides a guide to the design of a fire protection infrastructure. Appendix B, A Resilience Analysis of Metrolink 111 provides an analysis of the September 12, 2008 Metrolink 111 accident near Los Angeles. This analysis includes the contributions to the accident of all the nodes of the transportation infrastructure system of systems associated with that accident. This accident is also discussed in Chapter 4, Case Histories.

Figure 2.3 depicts a product-centered infrastructure as described above. This view shows how the various "nodes" of this infrastructure interact with each other. This interaction is key to the integration of an infrastructure system and to system resilience.

Both views show the following features:

- Nodes of the system. For Figure 2.2 and the infrastructure system, it shows all the nodes of the infrastructure. For Figure 2.3 and the product-centered infrastructure, it shows the various nodes associated with the developer, customer and suppliers.
- The activities performed by each node. For example, the activity of the transportation node in Figure 2.2 is "move people."
- The relations among the nodes. For example the relation between the power node and the telecommunications is to provide power.

7.1.2 Organizational View

The organizational view is primarily an internal view of an organization, especially an organization that is either a stand-alone human system or a node of a larger system, such as a civil infrastructure system of systems. This organization does have external links, and these links are pertinent to achieving resilience.

However, an organizational structure may have an effect on several aspects of system resilience. These factors apply to all categories of human-intensive systems, such as civil infrastructure systems, health care systems and product-centered systems. The two subjects of interest are interelement collaboration and authority.

7.1.2.1 Interelement Collaboration. Chapter 2, System Resilience and Related Concepts, lists four attributes of resilience which includes interelement collaboration. Interelement collaboration pertains to communication and cooperation of all elements of a system whether they may be external or internal. This collaboration can be either horizontal or vertical. That is, it can pertain to the relations among various organizations at any level of the organizational structure; it can pertain to the relations among organizations at various levels of this structure.

Organizational structure and policy can have an effect on this communication. If the safety organization, for example, in either a civil infrastructure or product-centered infrastructure is only one organizational level below the organizational lead person, it is only logical that communication between the safety organization and the leader would be superior to, say, six levels. Communication across multiple organizational tiers is inherently more difficult.

Leveson (1995) reinforces the view that organization is critical to safety by pointing to the following three deficiencies with an organizational focus:

- Responsibilities for safety responsibilities and authority are often diffused.
- Safety organizations often lack independence and have a low level status.
- Information channels and communications are often poor.

Next, there is the issue of "stovepipes" which are sometimes called "silos." This is also a communication issue. Stovepipes are organizational units that for cultural or personal reasons refuse to work with or exchange information with other organizational units. The net result is that communication and cooperation may be deficient. Although stovepipes are almost universally acknowledged to be detrimental to organizational effectiveness, methods of dealing with stovepipes have not been so easily identified. The concept of the integrated product team (IPT) discussed below is one way.

7.1.2.2 Authority. The second factor is authority. In most organizations, higher level organizations have decision-making authority over lower tier subordinate organizations. For this reason, the system resilience critical organizations will be more effective if they are in a position of authority, that is to say, at a higher organizational tier.

Maxwell (2009) notes that the San Francisco Fire Department has a three-tier authority process. The firefighters are trained to be aggressive; mid-level officers are trained to be aggressive but logical; and incident commanders are trained to be cautious when overseeing a fire or other incident.

The following paragraphs will show how the organization can be structured to reduce the effects of issues.

7.2 THE ORGANIZATION AND LATENT CONDITIONS

Chapter 3, Disruptions, notes that accidents can be initiated by disruptions that have their origin in latent conditions. Latent conditions are organizational in nature. They are the long-standing deficiencies in an organization that eventually result in an accident. Examples given by Reason (1997) include gaps in supervision, poor design, and undetected manufacturing defects. Reason notes that these conditions may last for many years before they eventually result in an accident. Paté-Cornell (1990) documents the organizational factors that led to the North Sea oil platform disaster. It is the existence of these conditions that renders the organizational factors discussed in this chapter relevant.

7.3 THE IPT CONCEPT

The IPT concept applies primarily to product-centered infrastructures. However, some aspects of this concept may apply to, for example, civil infrastructures. One way that has been used to address the issues of communication and authority is the IPT concept shown notionally in Figure 7.1. Some aspects of IPTs are presented next.

First, the program structure matches the physical architecture of the product being designed. Figure 8.2 assumes that the product being designed is an aircraft. The program manager is the lead person for the entire aircraft. Hence, the organizational box for the program manager represents the entire aircraft.

Second, each design team under the program manager represents a piece of the physical architecture of the system. This is the *product* portion of IPT. This organizational structure allows each IPT to focus on a single element of the product system while bringing technical expertise from various disciplines.

Each of these design teams is an IPT. The "integrated" in IPT means that the team has all the people needed to bring the product into being.

In the study of resilience, the concept of the IPT has particular importance in the consideration of the issue of unanticipated interactions among components of a technological system. Chapter 8, Resilience Architecting, discusses how an IPT might have been brought to bear to resolve the interaction between the

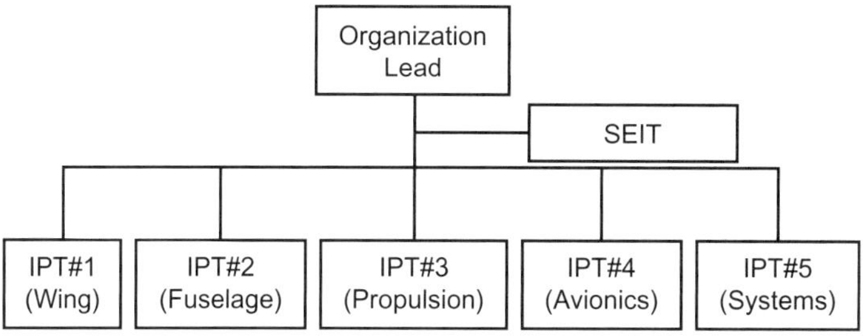

Figure 7.1. The integrated product team (IPT) concept.

strut and the software on the Mars Polar Lander case as discussed in Chapter 4, Case Histories. In this case, the structural engineers and the software engineers might have been able to identify the interaction before the system was built.

From a resilience point of view, the IPT structure often has a top-level element sometimes called the systems engineering and integration team (SEIT). This element is relevant to resilience because it contains key functions, such as safety. It is also conceivable that the organization may have a resilience function in the future. The SEIT is a logical home for the resilience function.

Leveson (1995) lists the following five requirements for the safety function within an organization:

- A direct link to decision makers
- Independence
- Direct communication with other parts of the organization
- Direct influence on decision making
- Focus and coordination

Of these five requirements, probably the most difficult to implement is independence. There are at least three logical ways to implement this requirement. First, the program manager can grant that independence to the safety organization. Of course, this approach requires some insight on the part of the manager. The second way is through the regulatory process. The Federal Aviation Administration (FAA), for example, designates certain employees of aircraft companies to be Designated Engineering Representatives (DERs). The DERs would wear two hats, figuratively, one for the FAA and one for the company. The third way is through independent reviews. These reviews are discussed in Chapter 9, Governance. Independent reviews would also require action on the part of the program manager.

The safety organization's position facilitates its role in resilience, especially the direct link, direct communications, and direct influence. The independence and focus requirements should be assigned to the organization by program management.

The IPT concept applies to nonproduct infrastructures as well. A firefighting infrastructure, for example, may be divided into various suborganizations, such as medical, fire suppression, and so on. However, assuming the infrastructure has a safety function and perhaps a resilience function, the position of these functions is most effective when positioned in accordance with the IPT structure.

7.4 INFRASTRUCTURE ARCHITECTING

Chapter 6, Capabilities, discusses the concept and discipline of systems architecting. A systems architect is needed to create an infrastructure. It can be recalled that systems architecting pertains to organizational systems as well as hardware and software systems. Rechtin (1991) explains how the systems architecting of organizations is done using heuristics. If a system, for example, and organizational systems contain many elements, for example, subgroups, the systems architect determines the "best" way to arrange these groups.

Because architecting is the discipline required to make the infrastructure resilient, a person with architecting skills should be a member of any infrastructure.

7.5 THE INFRASTRUCTURE AS A SYSTEM OF SYSTEMS

As we discussed in Chapter 2, System Resilience and Related Concepts, a system of systems is a combination of systems that act together to achieve a common objective. The obvious question is: When more than one system is working together, who is in charge? Who is the system architect?

The normal answer to this question is that all systems should collaborate. However, this answer is not completely satisfactory. There is an old saying that "when everyone is in charge, no one is in charge." Nevertheless, responsibility can be divided up as follows; we will take an emergency infrastructure as an example.

Individual node responsibilities—All nodes have some responsibility whether or not it is directed from above. Police, fire, and medical nodes, for example, have a responsibility to maintain communications with all other nodes.

Sub-group nodes—Provincial, county, or state entities have responsibilities for all nodes under them, such as individual police nodes. They also have the responsibility for determining and funding resources at lower levels. For example, how many fire trucks will be necessary at each individual node to provide assistance at a distant point? How many emergency rooms will be necessary to handle a pandemic? Systems architects at this level will need to answer these questions. The systems architects at this level can assure collaboration at lower levels.

Total-system node—This term simply means that some node of the system of systems has primary responsibility. This node can be either at the national or international level. Although the node at this level may not have the resources or authority to implement a system of systems, this node does have regulatory authority, analysis capability, and coordination authority. Its systems architects can determine, for example, how many emergency rooms are required on a national level.

Intermediate-system node—The idea here is that if one of the system-of-systems nodes cannot take a primary role, then an intermediate organization can. One example is the Global Earth Observation System of Systems (GEOSS) (2006). GEOSS is a consortium of societies and other organizations that endeavor to help governments around the world share data regarding natural disasters, such as earthquakes, tsunamis, and hurricanes. The set of governments is the system of systems, and GEOSS is the intermediate-system node. This role can be played by government agencies and international organizations, such as the United Nations, even when they do not have legislative authority to mandate actions by the individual nodes.

Another example of an intermediate system is the Commercial Aviation Safety Team (CAST) (2007). This organization is a cooperative effort of airline industries around the world to seek ways to reduce accidents in that industry.

Another example is a consortium of railroad companies who have cooperated to produce the General Code of Operating Rules (GCOR) (1994) for the railroad industry. The companies within this consortium have voluntarily adopted these rules.

An example of an intermediate system with a role in system resilience is The Infrastructure Security Partnership (TISP). TISP brings together industrial enterprises, government agencies, universities, and professional societies to find ways to make the civil infrastructure resilient to terrorist attacks and natural disasters.

So, systems architecting in a system of systems is not easy. Similar analyses can be made for space systems and commercial aircraft systems. The lesson is that collaboration is essential and possible.

7.6 NETWORK-BASED INFRASTRUCTURES

In the study of resilience, network-based infrastructures present unique vulnerabilities. Network-based infrastructures, as described by Garbin and Shortle (2007), carry commodities over long distances through multiple nodes. Examples are telecommunications, natural gas, oil, water, and electrical power. Garbin and Shortle describe these infrastructures in terms of four parameters shown in Table 7.1.

If a network is designed to carry a commodity, for example electricity, over a fixed set of links, then the network is brittle. The loss or failure of a single component may result in the failure of all or a major portion of the network. This kink of network system is called tightly coupled, as described in Chapter 2, System Resilience and Related Concepts.

Table 7.1. Network-Based Infrastructure Characteristics

Characteristic	Description
Demand	The amount of a commodity to be carried over the infrastructure. Associated parameters are throughput, loss, and delay.
Topology	The combination of the logical, and physical links of the network. The logical links are the map of where the commodity is sent from and to. The physical topology is the map of physical links.
Capacity	The actual amount of the commodity that the network can carry over specific links.
Routing	The determination of the logical links of the network.

Furthermore, according to Garbin and Shortle (2007), these brittle networks were found to be prohibitively costly to defend against the loss of a single node. This analysis was conducted using Cold War scenarios; however, it can be assumed that the cost of defending against a terrorist attack would yield similar results.

Referring to the major power grid failure in the northeast United States in 2003, Perelman (2007, p. 28) observes that "the sheer scale and complexity of modern power grids makes periodic, disastrous failures inevitable." Perelman adds that measures typically taken to protect such systems "tend to be ineffective or even to make future blackouts bigger and more likely." Perelman's comments reinforce the conclusion, discussed in Chapter 2, System Resilience and Related Concepts, that the concept of protection is less effective than resilience.

Another brittleness factor is the dependence of one network on the other. Garbin and Shortle (2007) cite the example of a telecommunications network that is dependent on the electrical grid for power.

The goal is to make the network flexible and tolerant, as described in Chapter 8, Resilience Architecting. This type of system would be called loosely coupled. There would be alternative ways to reroute the commodity in the event of a failure of a single node. So, from the point of view of resilience, the characteristic of routing in Table 7.1 carries a particular importance.

7.7 FURTHER EXPLORATION

1. Describe the infrastructure and draw the OV-2 diagram for several infrastructures. Suggested infrastructures are: a hospital, a hurricane response infrastructure, an anti-terrorism infrastructure, a space system infrastructure, a commercial aircraft infrastructure, a nuclear power infrastructure.
2. For a product system, describe and draw the internal infrastructure showing linking to outside entities.

Chapter **8**

Resilience Architecting

When we say that resilience can be architected, we mean that systems can be defined and the elements or the system can be arranged for which resilience will be an emergent property, that is to say, a property not possessed by the individual system elements. Maier and Rechtin (2009, p. 423) define an architecture as "the structure—in terms of components, connections and constraints—of a product, process, or element." Maier and Rechtin (2009, p. 423) also define architecting to be "the process of creating and building architectures. . . ." However, Maier and Rechtin reveal that architecting "may or may not be considered a separable part of engineering." In this book, the broader interpretation is used, that is to say, that architecting considers all aspects of defining the structure of a system.

System resilience results from a basic set of attributes that a system should have to be resilient. These attributes are traceable to a set of principles that have been developed and are, for the most part, accepted by the resilience community, as reflected in Hollnagel et al. (2006). These attributes are the source of design heuristics; they can be developed from the attributes.

8.1 HEURISTICS

The term *heuristic* means a design principle, guideline, or rule that has been learned from experience, especially with respect to the definition of the architecture of a system. Maier and Rechtin (2009, p. 424) define heuristic as "a guideline for architecting, engineering or design." This term is widely used among practitioners of systems architecting, such as Rechtin (1991).

Architecting Resilient Systems: Accident Avoidance and Survival and Recovery from Disruptions.
By Scott Jackson

Although certain case studies in Chapter 4, Case Histories, reveal inherent resilience, for example, Apollo 13 and the New York Power Restoration cases, the discipline has not reached a mature enough level in implementation to say that many systems have been built with resilience *intentionally* built in. Probably the one domain in which resilience is firmly established is aircraft cockpit design, where the Billings (1997) heuristics have been implemented for many years.

A notable proponent of the use of heuristics for architecting a system is Rechtin (1991). The attributes below will generate a set of heuristics that augment Rechtin's. When discussing adaptability, heuristics, and architecting, it is necessary to remember that we are entering the domain of experience, judgment, and intuition. Therefore, the results cannot be quantified or even validated except by experience, but not by traditional verification methods, such as testing: However, the architecture can be tested after requirements have been allocated and the system built.

So, if heuristics cannot be validated, what can we learn from them? The basic thing we can learn is how to structure a system before requirements can be allocated to the individual elements using traditional analytic methods. Second, we can learn that certain design attributes are necessary that go beyond the verifiable requirements and will yield value when they are employed. This value is reflected in the capability of being resilient to risks, cultural factors, and unknown threats. Of course, the selected design can be verified against known threats, but the expectation is that the system will be more capable than that.

8.2 THE CONTEXT OF HEURISTICS

The validity of a heuristic can be determined by four factors: the type of system being architected, the aspect of resilience being addressed, the type of disruption being encountered, and the domain of application. These factors will be discussed more thoroughly here.

8.2.1 Type of System

Heuristics depend on the type of system being architected; in the larger sense, that is some heuristics will apply to all systems including governments and infrastructure systems; others will apply to product-centered infrastructure systems; others will apply to technical systems; others will apply to technical systems with human components, such as pilots or maintainers; and others will apply to socioecological systems in which there is human intervention. Each of these system types is discussed in Chapter 2, System Resilience and Related Concepts.

8.2.2 Resilience Phase

In addition to system types, Chapter 2 describes the three aspects of resilience, namely, accident avoidance and survival and recovery from disruptions. So, when we ask the question of how to architect a resilient system, which of these three aspects are we referring to? The answer is perhaps all three or perhaps one or two. The following discussion will identify the resilience aspect that pertains to each way of architecting a resilient system.

8.2.3 Type of Disruption

The third factor that influences the selection of heuristics is the type of disruption being encountered. Chapter 3, Disruptions, describes two types of disruptions: first, degradation of input including unexpected changes in environment; and second, degradation of function, capability, or capacity. Of course, if the type of disturbance is not known, then the system may have to be designed to all types of systems. However, it is safe to say that certain types of systems will be more likely to encounter specific types of disturbances. For example, physical systems, such as buildings, dykes, and levees, are more likely to encounter unexpected changes in environmental load, although the other types of disturbances cannot be ruled out.

8.2.4 Application domains

Table 2.2 in Chapter 2, System Resilience and Related Concepts, provides a list of typical application domains for which heuristics may be applicable. That is to say, the heuristics for a civil infrastructure may be different from the heuristics for a commercial enterprise.

The discussion at the end of this chapter illustrates how heuristics complement each other in two ways: First, the system is more resilient even if all the heuristics cannot be implemented, and second, some heuristics are more effective if other heuristics are also implemented.

The following subsections treat three considerations in the architecting of resilient systems: adaptability, risk, and culture.

8.3 CONSIDERATION: ADAPTABILITY

One challenge of resilience is to determine how to survive a disruption of either of the two types defined in Chapter 3, Disruptions. That is the subject of adaptability. Adaptability applies to all the system types listed above, both product centered and human.

This capability is generally considered to be the most important of all capabilities. That is why it is being discussed separately from all the other capabilities mentioned in Chapter 6, Capabilities. In the end, adaptability is the primary defense against the disruptions discussed in Chapter 3, Disruptions. When disruptions cannot be predicted, systems need to be built that can survive despite the disruptions.

We discussed in Chapter 3 that disruptions have been found to result in accidents. So, the question is as follows: how do we design a system that is resilient to disruptions? The short answer is adaptability. The system needs to be adaptable to all possible disruptions and their possible consequences. Some systems have a significant amount of built-in adaptability. According to Westrum (2006b), the federal, state, and local infrastructure charged with dealing with hurricanes in New Orleans had little or no adaptability. Probably the best example of built-in adaptability is Apollo 13, which is described by Leveson (1995, pp. 558–563). In short, the crew, using its own ingenuity and advice from ground control, was able to *restructure* the system and use an auxiliary module (the lunar module) to survive. The ability to restructure is a key aspect of adaptability.

So, when we make a system adaptable, we are speaking primarily of the latter two aspects of resilience, namely, survival and recovery. Accident avoidance may be enhanced as well, but the factors that enhance survival and recovery will also enhance avoidance. One of Woods' (2007) basic heuristics, for example, is that a resilient system should be able to detect drifts toward failure and be able to make the decision to perform corrective action. So then, if this detection and decision-making capability can help a system avoid failure, then it could also help in survival and recovery.

Mendoça and Wallace (2006) provide an example of an incident with both adaptability and a lack of adaptability, that is, the restoration of power in Lower Manhattan on September 11, 2001 after the attacks on the two towers. In short, there was excellent cooperation among the four agencies on site: the U.S. Army, the New York police, the New York Fire Department, and the power company. They were able to restore power within 5 hours. However, a lack of ability to get fuel through the tunnels prevented them from maintaining that power indefinitely. Only through the intervention of the federal government were they able to do that.

An attribute is a characteristic or quality of something, for example, a system. The attributes of adaptability have been derived from logic and experience. Woods (2006b) has documented a set of adaptability principles from which these attributes emerge. The attributes listed in this chapter can be mapped to those principles.

This book pertains to both organizational, that is human, systems and to physical systems. Rechtin (2000) also made this distinction by writing a book on the architecting of organizational systems. The heuristics for organizations will, of course, be somewhat different from those for physical systems. The systems of interest for Hurricane Katrina as described by Westrum (2006b) and for the New York Power Restoration as discussed by Mendoça and Wallace (2006) were primarily organizational, whereas the Nagoya system was physical (an aircraft) as discussed by Leveson (2002) with a strong human interface, namely, the pilot.

A system with the following four attributes can be said to be resilient. These attributes can be architected into a system by use of the heuristics described for each one.

8.3.1 Attribute Number 1—Capacity

Capacity is the ability of a system to absorb or adapt to a disruption without a total loss of performance or structure.An example of capacity is the New York

Power Restoration of September 11, 2001. In this case, as described by Mendoça and Wallace (2006), the power company had a supply of portable generators in the event of a power loss. Obviously, the power company was not expecting a terrorist attack on the two towers. However, the system had a history of power losses, and this was another one, although it was significantly more severe. With these generators, they restored power within 5 hours.

It is probable that the generators were available because New York had a history of power outages. They had no history of terrorist attacks of this type. Nevertheless, this case shows that a reserve capacity built in for one purpose can be used for another purpose, namely, to recover from a terrorist attack.

This case also raises the question: Should the city of New York buy the portable generators even if they had a history of no power outages? The simple answer is yes, simply for the reason that the system is critical to the life of New York, and it needs to be resilient to all disruptions. One could argue that the city might not have had as many generators if they had not had a history of disruptions. This is true, and it illustrates, first, that the reserve capacity attribute is limited by the magnitude of possible disruptions. Second, it also illustrates that all the attributes should come into play. The attribute of interelement collaborations, which will be discussed below, can be used to assure that help might be received from remote locations.

Although attacks such as this are unpredictable, and resilience to all of them cannot be assured, the Mendoça and Wallace (2006) paper shows, first, that some systems contain a certain amount of natural resilience, and second when the attributes of adaptability are applied, a degree of resilience can be built into a system. From a national level, these attacks may have been predictable depending on the amount of intelligence available. However, even if they were predictable, resilience would then fall on the degree of interelement collaboration, that is, between the national and local authorities, which could have been brought to bear on the threat. The attribute of interelement collaboration is discussed later in this chapter.

The following sections are example heuristics that have been shown to enhance the buffering attribute.

8.3.1.1 The Absorption Heuristic. This heuristic, which is derived from Woods (2006b), states that the system should be capable of absorbing a disruption. This heuristic pertains to the strength of the system. It can be, on the one hand, the basic structural capacity to absorb a disruption. On the other hand, it can have to do with alternative ways to absorb a disruption. For example, a hurricane can be absorbed either by external means, such as seawalls, or by improved structural standards.

8.3.1.2 The Functional Redundancy Heuristic. This heuristic states that an alternative method should be available to perform each critical function that does not rely on the same physical systems. This heuristic, which is derived from logic and listed by Madni and Jackson (2009), simply says that the more ways there are to absorb a disruption, the more resilient the system will be. As discussed in Chapter 4, Case Histories, Apollo 13 was a good example of a system with multiple ways to survive and recover from a disruption.

The *functional redundancy* heuristic should not be confused with the physical redundancy heuristic below. The difference is that *physical redundancy* is the repetition of ways to perform the same function using physical systems, whereas the alternative heuristic calls for multiple ways to achieve the same goal using entirely different methods. For example, the citizens of a community struck by a hurricane may have different ways to survive and recover, either by remaining in the area and absorbing the disruption or by leaving.

8.3.1.3 The Physical Redundancy Heuristic. This heuristic states that physical redundancy should exist wherever possible. As discussed above and cited by Madni and Jackson (2009), physical redundancy is the ability to perform the same function in different ways. Physical redundancy employs redundant hardware and software to accomplish a goal when functional redundancy is not possible. Physical redundancy, as cited by Richards et al. (2007), is an accepted way to achieve increased reliability in contemporary engineering. It is an analytic approach whose numerical results can be calculated. The equations for reliability are well known and can be found, for example, in Blanchard and Fabrycky (2006, pp. 369–413). In technological systems, such as spacecraft and aircraft, physical redundancy is normally limited to lightweight components, such as electronics. Also, as discussed in Chapter 11, Cost, physical redundancy may be able to increase the reliability of a system with little added cost or weight.

8.3.1.4 The Margin Heuristic. The *margin* heuristic states that the system should have adequate margin to absorb disruptions. The design margin is a standard part of engineering, especially with respect to structures. Commercial aircraft structural design margins are typically around 50%. For the New Orleans levees, it is clear now that the margin for failure was entirely inadequate. It has been said that they were designed for the 100-year flood rather than the 1000-year flood. Margins themselves become eroded when systems age, as in the case of aircraft. Sometimes, the margins are inadequate because the external forces are unknown, as in the case of Hurricane Katrina. For Hurricane Katrina, this was a case in which two known elements would have lessened the impact of the storm. First, building the levees to the worst historical case and adding a margin to that level would have prevented the flooding. Second, implementing Woods' heuristic (2007) of observation-plus-decision would have enabled the population to take alternative steps, such as leaving the city. In this case, there was sufficient information, but no decisions were made to execute alternative steps. Finally, increased interelement collaboration to be discussed below would have given the population more options.

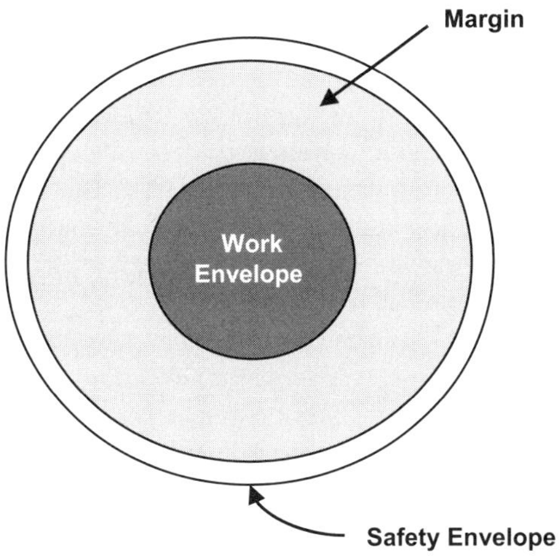

Figure 8.1. The concept of a work safety margin.

Barter and Morais (1998) apply the attribute of margin, not only to the physical system in a nuclear facility but also to a *work* system, that is, the system of activities that comprise experiments in the nuclear domain. Figure 8.1 illustrates the concept of margin in the work system.

With respect to margin, one needs to consider the penalties for margin. For example, an increased margin can result in both increased cost and reduced performance. An increased margin in aircraft or space system design can result in increased weight and cost of the system. These considerations will be the subject of design trade studies.

Schwartz (2008) notes that the National Aeronautics and Space Administration (NASA) margin of safety for human-rated launching systems is 40%. This margin, however, is against worst-case loads for *predicted* disruptions. It does not account for random disruptions, such as tiles that may impact the structure or O-rings that were required to operate outside the envelope of tested environments. For these reasons, the other heuristics should be employed.

8.3.1.5 The Hardness Heuristic. The *hardness* heuristic states that the system should be hard, that is, resistant to deformation. Cited by Richards et al. (2007), the *hardness* heuristic is particularly important for military technological systems that are threatened by Type A external disruptions. This reference mentions tank armor as an example. However, this heuristic would apply as well to many nontechnical systems that are threatened by disruptions from outside the system. The protection of astronauts from solar radiation is an example. One could say that hardening structures to earthquakes and hurricanes also fits this category. This heuristic applies to the survivability phase of resilience.

With respect to material hardness, it should be recognized that increased hardness can also result in increased brittleness, weight, and cost.

8.3.1.6 The Context-Spanning Heuristic. The *context-spanning* heuristic states that the system should be designed to both the worst case and most likely scenarios. It is often asked: What scenario should one design a system for? It is obviously impossible to design for all possible scenarios. A simple heuristic for this question comes from military planning, as discussed by Madni et al. (1987). The answer is that one should design to (1) the worst case scenario and (2) the most likely scenario. This answer does not satisfy all scenarios, but the answer provides a design that will be adaptable, robust and agile over the largest range of scenarios. This concept is depicted in Figure 8.2. One premise of the *context-spanning* heuristic is that if a system is resilient to these two predicted scenarios, then it will also be resilient to unpredicted scenarios.

Although it is impossible here to identify the worst-case scenario and the most likely scenario for all systems without some statistical and analytic data, it is possible to provide some hypothetical answers.

For a hurricane impact, it is possible to say that the worst-case scenario is most likely a category 5 hurricane hitting a major coastal city. The most likely scenario is probably a category 3 hurricane impacting a rural segment of the coast. Of course, it is possible for a category 3 hurricane to hit a city and for a category 5 hurricane to hit a nonurban segment of the coast. But the premise is that if authorities create codes and plans for both scenarios, then a set of adaptable, agile, and robust requirements will result.

As a second example, take a nuclear power plant. The worst-case scenario, of course, is an unstable reaction, such as in Chernobyl. The more likely scenario is a leak of radioactivity into a nearby ocean or river.

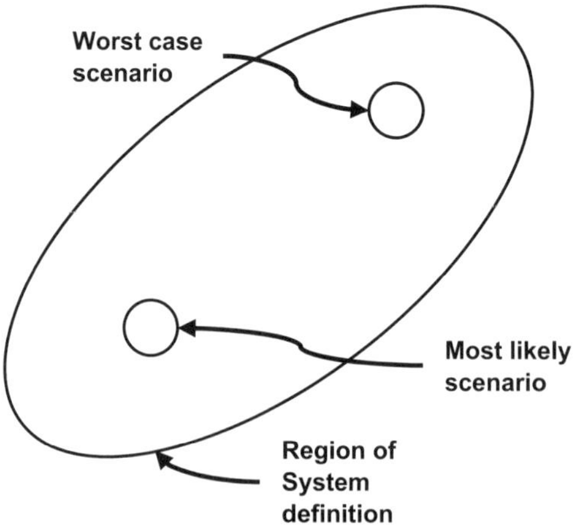

Figure 8.2. Scenario space.

In creating the worst-case and most likely scenarios, the case studies of Chapter 4, Case Histories, are of great value. For example, the fact that poor maintenance is a common root cause of catastrophes is of considerable value in establishing the most likely scenario, as in the case of Bhopal. This database of case histories is also replete with worst-case scenarios, such as those of Challenger and Chernobyl.

The case of U.S. Airways Flight 1549, which is also discussed in Chapter 4, illustrates the difficulty in designing to the worst case. In this case, the flock of birds exceeded the capacity of the aircraft engines to absorb them and continue operating. However, despite this limitation, the pilot of the aircraft was able to ditch the aircraft in water without loss of life, which illustrates the importance of the other resilience attributes present in that system, as discussed in Chapter 4.

With respect to context spanning, care should be taken to assure that the overall effectiveness is not sacrificed by sizing a system to one or the other of these scenarios. For example, designing to the worst-case scenario may result in a sacrifice in performance in more likely scenarios.

8.3.2 Attribute Number 2—Flexibility

Flexibility is a system's ability to restructure itself in response to disruptions. Westrum (2006a), for example, provides several examples of how important restructuring is to the survival and recovery of a system. He cites the tactics of football teams and military organizations as examples of restructuring. Examples of restructuring can also be found in nature. A colony of ants, for example, will restructure to cope with any disruption.

Following are some heuristics that are intended to enhance the attribute of flexibility, which includes the *reorganization* heuristic itself.

8.3.2.1 The Reorganization Heuristic. The *reorganization* heuristic states that the system should be able to restructure itself in response to disruptions or the anticipation of disruptions. Many existing and past systems had a large degree of natural, or built-in, restructuring ability. Apollo 13 was a good example. When the main power failed in the control module, the crew saved power by moving to the smaller landing module. In this way, they were restructuring the system, which is a key attribute of adaptability. It is not clear whether Apollo 13's adaptability was intentional or not; the key point was that it existed. Building in adaptability is a more difficult challenge. This heuristic is cited by Richards et al. (2007), who refer to it as system evolution.

An example of good restructuring after a major disruption was the New York Power Restoration as described by Mendoça and Wallace (2006) after the tower attacks on September 11, 2001. In this case, the New York power company was able to restructure and create a new organization to manage the system of portable generators that were available. This attribute gives rise to two potential heuristics, which are presented in the next sections.

8.3.2.2 The Human Back-Up Heuristic. The *human back-up* heuristic states that humans should be able to back up the automated system when there is a change in context the automated system cannot handle and there is time for human intervention. This heuristic is cited by Madni and Jackson (2009) and is implemented extensively in real-world systems. The reason is simple: Humans can analyze and respond to unpredictable disruptions. Chapter 1, On Resilience, discusses how humans are a paradox: that is, they can be both the source of disruptions and they can solve unexpected problems.

Take commercial aircraft, for example. Although operating a commercial aircraft completely automatically can be assumed to be within the current state of the art, removing human pilots from the system would deprive the system of the capability to deal with situations not anticipated by the designers of the avionics. It can be assumed, therefore, that humans will be present in the cockpit for the foreseeable future.

Leveson (1995, pp. 120–123) cautions that this heuristic is only applicable if the human is well trained and the automated system is designed to allow the human back-up.

8.3.2.3 The Human-in-the-Loop Heuristic. This heuristic states that humans should be elements of the system when there is a need for human cognition. This heuristic is cited by Madni and Jackson (2009). Probably the best example of the *human-in-the-loop* heuristic is the air traffic control system, which relies almost entirely on human cognition with extensive technological support. In this case, the human is not a back-up element but rather the primary decision-making element.

8.3.2.4 The Diversity Heuristic. This heuristic states that diversity should exist within systems. This heuristic is cited most often for socioecological systems; however, it also has applicability for other system types as well, especially technological systems.

Because socioecological systems are subject to human intervention, they are a subject of system resilience. Socioecological systems are examples of the macrosystems listed at the beginning of this chapter.

According to the Resilience Alliance (2008), the key to resilience in social-ecological systems is diversity. Biodiversity plays a crucial role by providing functional redundancy. For example, in a grassland ecosystem, several different species will commonly perform nitrogen fixation, but each species may respond differently to climatic events, thus ensuring that even though some species may be lost, the process of nitrogen fixation within the grassland ecosystem will continue. Similarly, when the management of a resource is shared by a diverse group of stakeholders (e.g., local resource users, research scientists, community members with traditional knowledge, government representatives, etc.), decision making is better informed and more options exist for testing policies. Active adaptive management whereby management actions are designed as experiments encourages learning and novelty, thus increasing resilience in social-ecological systems.

As noted by Richards et al. (2007), technological systems, such as commercial aircraft, also employ diversity as a design practice. For example, an aircraft might have both electrical and mechanical control systems for diversity.

8.3.2.5 The Predictability Heuristic. This heuristic states that automated systems should behave predictably and allow human override. Another Billings (1997) heuristic, this may be a difficult rule to comply with because it implies that the designer of the automated system must predict everything that the automated system might do. At least the automated system actions should be known to other elements in the system. In the case of the Mars Polar Lander, as described by Leveson (2002), not even the designer of the system knew what the software was going to do. In the case of Nagoya as described by Zarboutis and Wright (2006), obviously the pilot did not know what the flight control system would do. In the case of Chernobyl as described, for example, by Chiles (2002, pp. 162–163), the operators did not understand the unpredictable nature of the reactor control system.

Hardman (2008) calls this the *consistency* heuristic. He says that the human should be able to depend on platform conventions. Hardman suggests that standardization is one way to achieve this.

A subheuristic of the *predictability* heuristic is the *human-in-control heuristic—The human operator should be in command.* We have said before that the human is the primary contributor to unpredicted disruptions, as discussed in Chapter 3, Disruptions. It is ironic then that the human is the primary mechanism for adaptability. Why is this so? It is because the human has the capacity for sizing up the situation and recognizing options that the programmers of software could never have anticipated. Billings (1997) calls this rule an "axiom," which means that its truth is self-evident.

There is the danger, however, that this heuristic might be misinterpreted as evidence of the so-called "white scarf" syndrome. The "white scarf" syndrome takes its name from the romantic image of pilots in the early days of aviation who wore white scarves when they flew. The syndrome has come to imply the belief that humans are always better than machines. The *human-in-control* heuristic does not make that assumption. It merely says that when there is a situation that the pilot can assess better than the machine, the pilot should make that determination and the final decision.

Hardman (2008) calls this heuristic the *user control* heuristic. The key aspect cited by Hardman is the user's ability to undo and redo all processes within the system's control.

At the risk of repetition, Nagoya as described by Leveson (2002) and Apollo 13 also described by Leveson (1995, pp. 558–563) provide examples of both good adaptability and the lack of adaptability because of the involvement, or lack of involvement, of humans. In the case of Nagoya, the operator, that is, the pilot, was not in command. The flight control system had the final authority over whether to land or go around. If the pilot had had that authority, then the outcome might have been different.

It is fair to say that sometimes the automated system knows more than the operator. That is to say, the automated system may prevent the operator from doing something that is dangerous, such as exceeding the load limit of the aircraft. However, on balance, the heuristic says that if the pilot makes that decision, it will be correct more often than it is wrong.

In the case of Apollo 13, as described by Leveson (1995, pp. 558–563), it is not apparent that the spacecraft was ever designed to allow the crew to use the landing module if the power were to be lost. However, the fact is that the crew, in communication with the ground, made the right decision completely.

8.3.2.6 The Inspectability Heuristic. This heuristic states that the system should enable humans to take actions when needed without making unsubstantiated assumptions. This heuristic is cited by Madni (2007). Once again, the Nagoya accident, as discussed in Chapter 4, Case Histories, is a good example of a case in which this heuristic was needed.

8.3.2.7 The Simplicity Heuristic. This rule, which is cited by Billings (1997), may be a case of stating the obvious. It states that automated systems should be simple to train, learn, and operate. However, making automation simple to learn and operate is becoming increasingly challenging in the day of computer technology.

Hardman (2008, p. 21) also cites the *help and documentation* heuristic, which is another corollary of the *simplicity* heuristic. In this heuristic, "the information should be easy to search, focused on the present context, and list concrete steps to be carried out."

8.3.2.8 The Complexity Avoidance Heuristic. This heuristic states that complexity should only reflect the complexity demanded by the system functionality. Cited by Madni and Jackson (2009), this heuristic simply calls for good design practices that would use, for example, the minimum number of system elements and the minimum number of interfaces. This is a very old heuristic and has been stated in many ways. For example, Rechtin (1991, p. 19) gives three different versions. First, there is the keep-it-simple, stupid (KISS) heuristic. Another version states it as simplify, simplify, simplify. Finally, Rechtin cites Occam's Razor, which states that the simplest solution is usually the correct one. Actually Occam's Razor has a more philosophical origin: The basis of this observation is that nature generally seeks the simplest solutions. However, this observation does not preclude Occam's Razor as the basis for a heuristic.

For human-machine systems, Hardman (2008) cites the *minimalist design* heuristic. In this heuristic, all necessary information is available, supplementary information is retrievable, and tasks are simplified to relieve user memory load.

8.3.2.9 The Loose Coupling Heuristic. This heuristic states that the system should be repairable. This heuristic cited by Richards et al. (2007) is probably intended primarily for technological systems, in particular, military systems, in which damage is expected. Of course, repairing a system implies that there is time available to perform repairs. Therefore, it is compatible with the *neutral state* heuristic discussed below. However, this heuristic can be applied to any system for which time is available.

Spacecraft are among the most difficult systems to repair, but, as the world observed, the Hubble satellite was eventually repaired and able to continue its mission. The repair of civil infrastructure systems, such as power systems, will be dependent on the number of spare parts and other material that are damaged. Richards et al. (2007) also list replacement of damaged elements and cite cases in which satellites have been replaced.

Once again, caution should be exercised with respect to this heuristic. This heuristic can only be implemented if it is concluded that a system can be repaired without sacrificing other system qualities, such as reliability.

8.3.2.10 The Loose Coupling Heuristic. This heuristic states that the system should allow for flexibility in processes and decisions. This heuristic is based on the observations of Grote (2006) and Perrow (1984, pp. 92, 96), who point to the importance of rules in an organization. According to Grote, three kinds of rules are goal rules, process rules, and action rules. Among these, process rules and action rules are the least flexible with respect to organizational decisions. Goal rules allow for alternative methods and resources to be used in any situation, even situations that were not foreseen. This principle is known as loose coupling as discussed in Chapter 2, System Resilience and Related Concepts.

8.3.3 Attribute Number 3–Tolerance

Adaptability requires that a system be tolerant to disruptions, that is, it will exhibit "graceful degradation" near the boundary of its performance. Probably the best example of a system lacking tolerance was Chernobyl, which was discussed in Chapter 4, Case Histories. Although Chernobyl did not fail because of component malfunction, its lack of tolerance to human input caused it to go into a highly unstable mode and fail.

Woods and Cook (2006) emphasize that performance relative to a boundary is not a static quality but rather is dynamic. They point to automated response systems, for example, on aircraft that cannot compensate for a disturbance when the system gets near its boundary. This phenomenon is called "decompensation." They recommend that adaptive capacity be monitored continuously, and they intervene to increase it. They also recommend studying case histories to determine the boundaries of adaptive capacity.

In the design of any system, Woods (2007) says that adaptability requires observability plus decision. By this he means that you must know what is happening and be able to change the path that may lead to disaster. This rule is reflected in *drift correction* heuristic, which is discussed below. Again, the Mars Polar Lander, as discussed in Chapter 4, is an example. The idea is that if instrumentation had been available, for example telemetry, it would have given the mission controllers information about the interaction between the struts and the software, and they could have take action to correct the situation.

8.3.3.1 The Graceful Degradation Heuristic. This heuristic states that the system should degrade gradually when exposed to a disruption. We have discussed cases in which failure was not immediate and catastrophic and some that were. Columbia, for example, was immediate and catastrophic. Graceful degradation is sometimes known as "soft" failure. The Sioux City DC-10 accident was, however, an example of graceful degradation. As discussed in Chapter 4, Case Histories, this aircraft managed to land and to limit fatalities through the means of propulsion control. This heuristic is cited by Woods (2006b).

Gribble (2001) argues that the interaction among components is the basic cause of system fragility and that graceful degradation is the only way to deal with it.

One way to achieve graceful degradation is through functional redundancy, as was observed in the case of the Sioux City DC-10 case study.

Another way to achieve graceful degradation is by localizing capability. A simple example of creating graceful degradation in the military domain is the transition from trenches in World War I to foxholes in World War II. A trench system is vulnerable to multiple casualties from a single artillery shell; however, a foxhole is vulnerable to only a single casualty per shell, which allows the balance of the fighting unit to survive and continue fighting.

This example can be repeated in many domains. In today's world, many power users have generators to provide power to a specific location. Hospitals are one example. Although the U.S. electrical grid system has many advantages, for example, the ability to share power among locations, it has one disadvantage, which is the vulnerability to network failures. One alternative is to have many localized back-up power systems. The New York Power Restoration case in Chapter 4 showed that back-up generators do provide graceful degradation and enhance resilience.

8.3.3.2 The Drift Correction Heuristic. This heuristic states that drift toward brittleness should be detected and corrected. Drift, which is defined by Woods (2006b), is a key idea in system resilience. That is to say, it is always better to know when catastrophe is approaching. The conventional ways to detect drift are through human observation, cameras, or other sensors. If drift is detected, then two options are available. First, corrective action can be taken to avoid the accident, or second, the system can be put into neutral to decide what the next step is, as described below in the *neutral state* heuristic.

The key factor in corrective action is time. There has to be time in which to make a *corrective action*. The *drift correction* heuristic is implied in Woods' (2006b) observability plus decision rule.

Corrective actions can be made either by humans, machines, or humans in conjunction with machines. Hardman (2008), for example, cites the *recognition and recovery from errors* heuristic, which can be considered a corollary to the *drift correction* heuristic. He says that automated systems should be capable of identifying problems and suggesting solutions.

The classic case of a missed opportunity for drift detection and correction was for Columbia as described by the Columbia Accident Investigation Board report (2003). According to the report, NASA was offered imagery support from the Department of Defense. NASA turned this down as a low-priority effort. This imagery might have provided the information to analyze the exterior damage to the vehicle and ultimately to repair it to avoid its catastrophic ending.

8.3.3.3 The Neutral State Heuristic. This heuristic, which is cited by Madni, et al. (1987) states that the system should be put into neutral if possible following a disruption. This heuristic implies a default capability that is intended to give the system the time to make decisions after an initial disruption has been experienced. Automated systems, for example, may have a "safe" mode that prevents it from shutting down or failing while it is being diagnosed.

According to Maxwell (2009), the San Francisco Fire Department practices the *neutral state* heuristic by training its firefighters to evaluate the situation before taking any risky actions, such as entering burning buildings.

8.3.3.4 The Automatic Function Heuristic. This heuristic states that functions should be automatic only if there is a good reason for making it automatic. This heuristic, as cited by Billings (1997), is based on the simple idea that the human is more adaptable than the machine and can make decisions that the machine is incapable of doing. Of course, the human should be operating under a set of workable rules that are easy to understand, and second, the operator is well-trained in the tasks. This attribute is in agreement with Grote's principles for rule-making described below.

Once again, this heuristic has the danger of being misinterpreted. It should not be read to mean that manual operation is always better than automatic operation. All it says is that if the situation is such that a function can be executed manually, then manual operation would be preferred in those situations.

8.3.3.5 The Organizational Decision-Making Heuristic. This heuristic states that organizational decision making should be monitored. This heuristic is most applicable to infrastructure systems of all types, which include both civil infrastructure systems and product-centered infrastructure systems. In general, there does not seem to be much doubt about the wisdom of this heuristic. That is why such practices as design reviews, peer reviews, and independent reviews, as discussed in Chapter 9, Governance, have become so common. All these techniques have been developed to counter what Reason (1997, pp. 10–11) calls *latent* conditions, that is, the deficiencies that exist in organizations that contribute to accidents at some date, perhaps years, before the actual accident. These conditions are discussed in Chapter 3, Disruptions, and in Chapter 7, Infrastructure.

The apparent assumption behind this heuristic is that individuals, on their own, will make flawed decisions, either out of a lack of expertise or other reasons, such as the conflicting-priorities paradigm discussed in Chapter 5, Culture. Reason (1997), for example, points out that flawed decisions are sometimes made with the best of intentions. Why, for example, would decisions be made to reduce maintenance in the cases of Bhopal, the Chicago 1979 disaster, the ValuJet accident, and the Alaska Airlines MD-80 crash, all of which are discussed in Chapter 4, Case Histories?

This heuristic can be implemented in many ways. However, it is generally agreed that oversight of all decisions is a necessary ingredient.

8.3.3.6 The Organizational Planning Heuristic. This heuristic states that signs should be noticed that call into question organizational models, plans, and routines. Also, applicable to all infrastructure systems, this heuristic calls into question the entire organizational process system. The Federal Aviation Administration (FAA) (2004), for example, identifies work environment standards as a key process area. This process area includes the consideration of all tools and processes needed to produce a product. This is another example of a heuristic that addresses the latent conditions, as defined by Reason (1997), that contribute to an accident long before it actually occurs.

The following five heuristics are cited by Richards et al. (2007) and apply primarily in the military domain. Hence, the disruptions involved are Type A, or external, threats encountered in that domain. In addition, the *regroup* heuristic was suggested by A. Raveh during open discussions of the Architecting Resilient Systems tutorial at the International Council on Systems Engineering (INCOSE) Symposium, Utrecht, the Netherlands, on June 16, 2008.

8.3.3.7 The Mobility Heuristic. This heuristic states that the system should be able to avoid a threat by moving. Also cited by Richards et al. (2007), this heuristic is particularly important in the military domain. Avoiding detection by moving is a common military tactic. This heuristic applies to the avoidance phase of resilience. Mobility can be used also to maneuver away from a threat. Richards et al. (2007) refer to this maneuver as avoidance.

8.3.3.8 The Prevention Heuristic. This heuristic states that the system should be able to suppress future potential disruptions. Cited by Richards et al. (2007), this heuristic is applicable to many domains, both military and nonmilitary. It is applicable to the avoidance phase of resilience and to external, or Type A, disruptions. Examples include efforts to prevent future terrorist attacks and prevent worldwide pandemics, such as the H1N1 virus.

8.3.3.9 The Retaliation Heuristic. This heuristic states that the system should be able to retaliate to a threat. Cited by Richards et al. (2007), this heuristic also applies primarily to the military domain. It can apply either to technological or human systems. It applies to the avoidance phase of resilience.

8.3.3.10 The Concealment Heuristic. This heuristic states that the system should attempt to conceal itself against potential threats. Another Richards et al. (2007) heuristic, although this rule is used widely in military tactics, such as for stealth bombers, it can be observed in nature in the efforts by animals, such as the chameleon, to conceal themselves. However, it can be said that it applies almost exclusively to situations in which the Type A, or external threat, is hostile.

8.3.3.11 The Deterrence Heuristic. This heuristic states that the system should attempt to deter hostile threats from attacking. This heuristic applies almost exclusively to the military domain and would be used during the avoidance phase of resilience against Type A, or external, hostile threats. This heuristic is cited by Richards et al. (2007).

8.3.3.12 The Regroup Heuristic. This heuristic states that the system should be able to restructure itself after a disruption to recover some degree of functionality and performance. This heuristic, which was suggested by Raveh (open discussion, INCOSE symposium, 2008, Utrecht, the Netherlands), pertains primarily to systems of systems or to systems with a large number of independently controlled elements, such as military personnel. This heuristic is seen in nature to a great extent. For example, a colony of ants regroups itself after being subjected to any outside disturbance, such as an attack by a predator.

In a civil infrastructure environment, this heuristic would be most useful to law-enforcement agencies and firefighting personnel.

8.3.4 Attribute Number 4—Interelement Collaboration

Another attribute of adaptability is interelement collaborations, or communication and cooperation between elements. Interelement collaborations in a human system context lead to local problem solving and how local adaptation can influence strategic goals and interactions. Hurricane Katrina, as discussed by Westrum (2006b), is an example in which there was virtually no communication or cooperation among government agencies. This system failed the adaptability test with catastrophic results. Apollo 13, as discussed by Leveson (1995, pp. 558–563), also used interelement collaboration for communication with the ground to survive.

Another example of a lack of interelement collaboration was the New York Power Restoration as described by Mendoça and Wallace (2006). In this case, although the city of New York power infrastructure after the tower attacks on September 11, 2001 exhibited significant adaptability in restructuring and margin, the lack of communication with the New Jersey government made it difficult to restore power in Manhattan.

Interelement collaboration is particularly important in the infrastructure system integrity, which is critical to system resilience. An example is the ValuJet accident discussed by Chiles (2002, pp. 150–155, 310). This example is important because interelement collaboration is normally not of exceptional quality across organizational boundaries of most infrastructures. Although Chiles gives a complete account of the accident, the root cause boiled down to some improperly loaded oxygen tanks without safety cap protectors. This fact was compounded with other root causes of lax regulation by the FAA and cost and schedule pressures among the root causes listed in Chapter 4, Case Histories.

But the important aspect of the ValuJet incident, from an interelement collaboration point of view, is that operational infrastructures, such as this one, normally contain many organizational interfaces, each of which is vulnerable to communications gaps and errors. In the case of ValuJet, there were five major interfaces that were important as shown in Figure 4.1. Apart from the regulatory aspects discussed by Chiles (2002), the important thing is that between the manufacturer and the maintenance division, or in the ValuJet case, an independent contractor, two interfaces contributed to the communications breakdown. So, given the number of interfaces to deal with, it is not surprising that such mishaps are common.

The ValuJet accident can also be called a *network failure*, because a failure in one node of the system, the aircraft support system, resulted in a failure of the whole system.

8.3.4.1 The Informed Operator Heuristic. This heuristic states that the human operator should be informed regarding all aspects of the system operation. If the previous heuristic about the human being in command is to be followed, then the human should know what the situation is, including what the automated system is doing or intends to do. In the case of Nagoya, as discussed by Leveson (2002), obviously the pilot was not informed. In the case of Apollo 13, also discussed by Leveson (1995, pp. 558–563), the crew was fully informed of the situation.

Hardman (2008) refers to this heuristic as the *feedback* heuristic and adds that the automated system should indicate the status of tasks in progress and indicate closure of the task.

8.3.4.2 The Hidden Interaction Heuristic. This heuristic states that hidden interactions among nodes of a system should be avoided. Cited by Woods (2006b) and Perrow (1984, p. 79), the unexpected interaction between components is a contributing factor in accidents. An early reference to this phenomenon is Perrow (1984, p. 364), who refers to these interactions as "unknown and unintended interactions." Chapter 3, Disruptions, discusses two-agent disruptions, that is, disruptions caused by the interaction between components. Chapter 3 shows how undesirable interactions can be separated from desirable interactions in an interface analysis that is an elaboration on a traditional interface analysis.

This is one of the subtlest and most difficult to implement heuristics because the interactions can result from many causes and finding these interactions during development is difficult. Leveson (2002), for example, notes that treatment of the interaction of properly operating components is not a standard part of traditional safety analysis. Although this heuristic may seem to apply primarily to technological products, it can also apply to human-intensive and other systems.

A notable technological example is the Mars Polar Lander, as discussed in Chapter 4, Case Histories. In this case, the vibration from a landing strut caused the engine to shut down prematurely. Both the landing strut and the software commanding the shutdown performed as designed.

The treatment of the Mars Polar Lander case can be performed either as a technological problem or as a management problem. As a technological problem, one way to anticipate this interaction is through detailed simulations of the hardware and software on the system. These types of simulations have become common in recent years. However, they are complex and expensive to develop and implement. Epstein (2006), for example, applies these types of simulations in the nuclear industry.

The second way to treat the hidden interaction of a technological system, such as the Mars Polar Lander, is to focus on the technical management aspect, that is, the product-centered system. Both Chapters 7, Infrastructure, and 9, Governance, provide ways to address the risk of hidden interactions. Chapter 7, for example, discusses the concept of the integrated product team (IPT). The idea is that products are developed with specialists from various disciplines who work together on a product, such as the Mars Polar Lander module. The concept is that if the structural engineers who designed the strut and the software engineers had worked together, this interaction might have been discussed and treated. Chapter 9, Governance, also provides other ways to ferret out hidden interactions. These include the independent review and the peer review.

Chapter 3, Disruptions, discusses using the multiple-agent N^2 diagram as a method for identifying both desirable and undesirable interactions among system components. This methodology can be used during the governance process as discussed in Chapter 9.

Hidden interactions also occur in human-intensive systems and systems of systems. In these cases, it is unlikely that simulations exist for analyzing interactions, but they may. In the case of the ValuJet accident, for example, one could say that the action of the airline support contractor of putting oxygen tanks on the aircraft was an unanticipated interaction. This is an example of a system of systems that consisted of the aircraft manufacturer, the airline, and the contractor. The prevention of such interactions would require the oversight of higher authorities, such as the FAA, the manufacturer, or the airline itself.

Finally, this heuristic demonstrates that it is just as important to deal with the disruption as it is to deal with how the system performs after the disruption occurs. So, if the disruption is avoided, then a catastrophic failure will also be avoided.

8.3.4.3 The Knowledge Between Nodes Heuristic. This heuristic states that knowledge between the nodes of a system should be maximized. This heuristic can be logically derived from the interelement collaboration attribute. It is a broader statement of other heuristics, for example, the *intent awareness* heuristic, which is discussed below. However, its usefulness is most apparent in the cases involving human systems-of-systems, particularly civil infrastructure systems. As described in Chapter 4, Case Histories, there was an absence of knowledge between nodes of the New Orleans civil infrastructure. In the case of the New York Power Restoration, there was considerably more knowledge between nodes.

8.3.4.4 The Human Monitoring Heuristic. This heuristic states that the automated system should be able to monitor the human operator. As described by Billings (1997), when an operator gives a command to an automated system, does the machine understand the operator's situation? In the Nagoya incident, did the flight control system know that the pilot wanted to land? The answer was obviously no, and this was the crux of the disconnect between the pilot and the flight control system.

8.3.4.5 The Automated System Monitoring Heuristic. This heuristic states that the human operator should be able to monitor the automated systems. This heuristic is a subset of the previous heuristic about being informed. Automated systems, for example, flight control systems, should provide the human operator sufficient information for the operator to be in control. In the case of Nagoya, as described by Leveson (2002), the operator was not able to monitor the flight control system completely.

In a separate heuristic, Hardman (2008, p. 20) refers to the *real-world agreement* heuristic, which calls for the user's language rather than computer-oriented terms. It also calls for the computer to provide information that does not conflict with the user's mental model.

Hardman (2008, p. 20) also cites the *error-prevention* heuristic, which supports the *automated system monitoring* heuristic. This heuristic calls for the automated system to provide confirmation before dangerous actions are taken.

8.3.4.6 The Intent Awareness Heuristic. This heuristic states that each element of the system should have knowledge of the others' intent and should back up each other when called on. Both Madni (2007) and Billings (1997) apply this rule to human automation. However, this rule could apply even to systems in which there are no humans. In the Mars Polar Lander, for example, the struts and the software were completely at odds on the intent of the propulsion shut down logic in the software.

This rule comes to the fore most strongly in both the case of the Mars Polar Lander, as described by Leveson (2002), and Nagoya, as described by Zarboutis and Wright (2006). In both cases, as described below in 8.4, Adaptability in Software, if the software had had knowledge of other elements' intent, it might have been able to be programmed to adjust for the non-normal conditions.

8.3.4.7 The Interelement Impediment Heuristic. This heuristic states that there should be no impediments to interelement collaborations. In the civil infrastructure domain, impediments to interelement collaboration have been the source of system brittleness. According to Jackson (2008), procedural impediments prevented the U.S. Marines from providing evacuation helicopters during the Southern California wildfires of 2007. Stephan (2007) notes that in the Hurricane Katrina disaster, the different elements of the New Orleans civic infrastructure could not communicate with each other because of the lack of *interoperability* of the communications equipment.

Nolte (2008) points to the difficulty in consolidating intelligence information from different sources in a terrorist environment. Nolte recommends a board to oversee this consolidation and hence eliminate this obstacle to interelement collaboration among agencies.

8.4 ADAPTABILITY IN SOFTWARE

Software presents a unique challenge in technological systems. The term *adaptability* is used most often in the software domain. However, this term is rarely used with respect to a system that consists primarily of hardware and humans. A typical example is a traffic intersection for which the light signals are controlled by a software module. The concept is that the traffic will be monitored by sensors at the intersection and that information will be used by the software to *adapt* to the changing traffic patterns and to adjust the signals to those patterns. A defense Advanced Research Projects Agency (ARPA) definition of self-adaptive software quoted by Laddaga and Robertson (n.d., p.1) is as follows: "Self-adaptive software evaluates its own behavior and changes behavior when the evaluation indicates that it is not accomplishing what the software is intended to do, or when better functionality or performance is possible." The traffic example, above, is one example of this capability. The software employs feedback loops to accomplish this adaptation.

Software adaptability is well within the scope of system resilience as defined in this book. For example, the software in the case of the Mars Polar Lander described by Leveson (2002) could have been designed to detect any off-normal conditions and adjust to them. Second, the software in the case of the Nagoya incident described by Zarboutis and Wright (2006) could have been designed to detect that the software and the pilot were in conflict and to defer to the pilot or adjust in some other way to that fact. This design would enable the system to comply with Billings' rule, above, that each element of the system should have knowledge of the others' intent.

8.5 LAWS

The term "law" is often used with respect to some attributes because they are considered always to be true under any circumstances. Although some of these laws were developed for software systems, many of them can be said to be true for any type of system.

8.5.1 Ashby's Law

One example of a commonly used law is Ashby's Law. It states, "The larger the variety of actions available to a control system, the larger the variety of perturbations it is able to compensate." In the terminology of this book, this law can be rephrased as, "The more ways you have to make a system resilient, the more resilient it will be." The heuristic that can be derived from this law is, "Provide as many opportunities for accident avoidance, survival and recovery as you can." This is just one example of a law that can be used in the definition of systems with resilience. Woods (2006c), in which Ashby's Law appears, is a good source for a collection of such laws.

8.6 CONSIDERATION: RISK

Risk management, as identified in Chapter 6, Capabilities, is a critical capability to accomplish system resilience. Risk management is a key factor in the first system resilience aspect, accident avoidance. Of course, if risk management is successful in achieving risk avoidance, then the other two aspects, survival and recovery, are moot, that is to say, they are automatically achieved. However, if the risk is mitigated, that is to say, reduced to a lower level, then survival and recovery are still at risk.

Risk management is the science of managing the possible consequences of *known* phenomena. Hence, it is useful in analyzing the possibility of events similar to those that have happened before, as documented in Chapter 4, Case Histories. Second, it can analyze phenomena that are highly unlikely but possible. This category can describe any disruptions described in Chapter 3, Disruptions. A valuable input to risk analysis is the results of the resilience prediction methods discussed in Chapter 10, Measuring Resilience. Hence, if measures of resilience can be found, and the evidence is that they do exist, then proactive steps can be taken to handle the risks associated with the lack of resilience, which is sometimes called brittleness.

Risk is not a static characteristic. It should be measured and evaluated continuously. Dekker (2006), for example, warns against "drifting" into a state of risk, that is, low resilience or brittleness. It can be said that a system that lacks adaptability will be in a state of risk because it will be vulnerable to threats it was otherwise not vulnerable to.

Leveson et al. (2006) show that risk can be modeled in a dynamic, nonlinear, continuous fashion. Leveson et al. asserts that their model is more complete than traditional risk models.

The shortcoming of risk management is that it is not structured to evaluate phenomena that have never happened before and are not possible to predict. The main defense against this type of disruption is adaptability, which is discussed above. On the one hand, this conclusion does imply that one cannot assess risks for unpredicted disturbances. However, it does not mean that risk analysis is useless, because, plenty of risks have happened before that need to be planned for. On the other hand, resilience promises to address disturbances that have never happened before because of the adaptability built into the system.

So, all the above having been said, what can we say about how risk should be treated? The following heuristics provide a beginning.

8.6.1 The *Small Problem Focus* Heuristic

This heuristic states that the focus of system development should be on small problems. We saw in Chapter 6, Capabilities, that most major accidents occur as a result of low-probability, high-consequence disruptions. This observation is contrary to the popular belief that one should focus on the high-likelihood disruptions. The law of large numbers, which was discussed in Chapter 3, Disruptions, concludes that if there are many opportunities for low-probability risks, then the probability of an accident will trend toward a fixed and stable value.

So, how can these low-probability events be identified? This is not an easy task, but methodologies are available. For technological systems, simulations are available that can identify the possible interactions among components. Rigorous management reviews can also identify the defects that might have gone unnoticed.

8.6.2 The *Risk Aggregation* Heuristic

This heuristic states that when multiple risks are being considered, the aggregation of risks should also be considered. It has been found in practice that risks are often considered independently. Vaughn (1996), for example, points out in the Challenger case that NASA at that time had a process for aggregating risks failed to put this process into practice.

8.6.3 The *Risk Culture* Heuristic

This heuristic states that when risks are being considered, the organizational culture should be open to and honest about the identification and treatment of risks. Here, we have a clear overlap in heuristics. The culture heuristics below provide guidance for either changing or containing deficient cultural attributes. However, among all cultural factors, risk aversion is among one of the most serious paradigms that are detrimental to resilience. Numerous case studies, for example, Challenger, Texas City—2005, and Bhopal are notable for the cultural deficiencies. These cases are discussed in Chapter 4, Case Histories.

8.7 CONSIDERATION: RESILIENCE PREDICTION

Chapter 10, Measuring System Resilience, shows that significant progress has been made in determining the degree of resilience from statistical analysis of the correlation between near misses, that is, small problems, and major accidents with fatalities. This correlation shows that there is a promise that resilience can be improved by measuring small problems and then making improvements to eliminate those problems, hence lowering the likelihood of major accidents. If such data can be obtained, then the results can be entered into a risk analysis, to manage the risks associated with the measuring resilience deficiency.

8.8 CONSIDERATION: CULTURE

As pointed out in Chapter 5, Culture, the Columbia Accident Investigation Board report (2003) concludes that culture was at least as important a factor in the Columbia accident as the technical factors. The same conclusion can be reached for many other accidents listed in Chapter 4, Case Histories.

So, how does one eliminate or reduce the cultural factors? The Columbia report does not provide any specific suggestions, only improvements in technical and managerial processes. It might be concluded, then, that the implication is that if these processes are improved, then the cultural factors will become moot.

Chapter 5 provides a list of methods to improve the cultural climate. In short, two approaches are available. First, the technical and managerial processes are improved to such an extent that the cultural attitudes do not matter. Second, the cultural factors can be addressed directly through training and workshops. Perhaps a combination of the two is required. Either way, this subject needs to be addressed.

So, what aspect of resilience do cultural initiatives address? It is obvious that it addresses the accident avoidance aspect, as the Columbia report suggests. However, addressing cultural aspects will improve the survival and recovery aspects because these aspects are not frequently addressed.

In short, treating culture in a resilience context can be treated in two ways, as reflected in the following heuristics.

8.8.1 The *Culture Management* Heuristic

This heuristic states that a system of oversight should be imposed throughout a system that will tend to negate cultural deficiencies. Although this method may do little to change culture, it is probably the more practical approach with more likely short-term results. Essentially, this approach is suggested in the Columbia and Texas City—2005 incidents as discussed in Chapter 4, Case Histories. Chapter 9, Governance, discusses how this oversight might be achieved. The oversight suggested in Chapter 9 is intended to be thorough, reviewing all aspects of any organization to the most detailed level.

8.8.2 The *Cultural Change* Heuristic

This heuristic states that initiatives should be implemented to change the culture and remove the defective aspects of that culture. This is the more difficult of the two culture heuristics. Chapter 5, Culture, provides a list of options of how this cultural change might be implemented. However, the supporting data on cultural change, especially in a resilience context, are scarce. There are many examples of cultural change to improve business objectives, but the effectiveness of these methods to improve resilience is unknown. Senge (1999) provides some examples of cultural change from a business perspective but not from a resilience perspective.

8.9 THE ROLE OF RULES IN SYSTEMS WITH HUMAN COMPONENTS

A key question that arises when developing rules for humans in any system is to what degree should the rule be prescriptive, that is, should it tell the human exactly what to do? These questions arise when developing manuals, for example, for pilots or for aircraft maintainers or for personnel in an operating room. Grote (2006) and Perrow (1984, pp. 92, 96) address this question. Grote's conclusion is that rules should be "loosely coupled," that is, they should address the goals of the operation and they should stop short of specifying specific actions. The second conclusion is that the person executing the rules should be well trained.

Grote's (2006) rules are completely compatible with, first, Woods' (2006b) principles of adaptability, especially the attribute of flexibility. Second, Grote's rules are in agreement with Billings' (1997) attributes of human-machine interfaces.

Finally, these rules, however logical they may seem, are no guarantee that human errors will not occur or that unpredicted disruptions, as discussed in Chapter 3, Disruptions, will not result.

8.10 COMPLEMENTARITY OF HEURISTICS

It should not be concluded that if a system does not meet all the heuristics discussed in this chapter, it will lack resilience. There will always be situations in which all the heuristics cannot be met, either because the designers were unaware of the threat or because costs or other constraints prohibited their implementation.

For example, if the designers of the New York power system, as described in Chapter 4, Case Histories, had not had a history of power outages, then they may not have had the extra generators on hand to fulfill the extra *capacity* attribute described below. In this case, the attribute of interelement collaboration, as described below, would still provide a degree of resilience. Similarly, even though the levees in New Orleans were not built to an adequate height and with adequate strength prior to Hurricane Katrina, as described also in Chapter 4, interelement collaboration would provide some resilience here also.

Also, some heuristics aid other heuristics. For example, it is only a possibility to implement the *neutral state* heuristic if the system is not completely destroyed or if the *drift correction* heuristic is in place.

8.11 THE RESILIENCE ARCHITECTURAL HIERARCHY

Table 8.1 is a summary of resilience architecting methods. It depicts an architectural hierarchy of the following four levels of resilience architecting: characteristic, attribute, heuristic, and example. Characteristics represent the aspect of resilience desired. Attributes, laws, and constraints represent the basic truths that drive resilience. Heuristics are the lessons from experience that provide design guidance. Examples are specific instances for which the heuristics apply.

Table 8.1. Summary of Resilience Architecting Heuristics

Heuristic	Phase	Disruption	Principal Systems
Consideration—Adaptability			
Attribute—Capacity			
Absorption	2	A	All
Functional redundancy	2	B	All
Physical redundancy	2	B	T
Margin	2	A	All
Hardness	2	A	T
Context spanning	2	A	All
Attribute—Flexibility			
Reorganization		A	All
Human-in-the-loop	2, 3	A	H-T
Diversity	2	A, B	B-E, T

(*Continued*)

Table 8.1. (Continued)

Heuristic	Phase	Disruption	Principal Systems
Human-in-control	2, 3	A, B	H-T
Predictability	2, 3	A, B	H-T
Simplicity	2, 3	A, B	H-T
Complexity avoidance	1	A, B	T
Reparability	3	A, B	T
Loose coupling	1, 2, 3	A, B	H
Attribute—Tolerance			
Graceful degradation	2	A, B	All
Drift correction	1, 2	A, B	All
Neutral state	1, 2	A, B	All
Automatic function	1,2	A, B	H-T
Organizational decision-making	1, 2	A, B	H
Organizational planning	1, 2	A, B	H
Mobility	1	A	H,T
Prevention	1	A	H,T
Retaliation	2	A	MT
Concealment	1	A	MT
Deterrence	1	A	MT
Attribute—Interelement collaborations			
Informed operator	1, 2	A, B	H-T
Hidden interaction	1, 2	B	All
Knowledge between nodes	1–3	A, B	All
Human monitoring	1–3	A, B	H-T
Automated system monitoring	1–3	A, B	H-T
Intent awareness	1–3	A, B	All
Consideration—Risk			
Look for small problems	1	B	All
Aggregate risks	1	A, B	All
Consider culture	1	A, B	All
Consideration—Culture			
Contain culture	1	A, B	All
Change culture	1	A, B	All

Code:

T = technological

H-T = human-technological

B-E = bio-ecological

H = human

MT = military

8.12 FURTHER EXPLORATION

1. From your research or from your own mind, create several heuristics, other than the ones mentioned in this book. Show how they relate to the attributes of adaptability and agility to make a system, either a human or product system, more resilient.
2. Take any case history discussed in this book or elsewhere and suggest methods of making it more resilient through adaptability, agility, and robustness.
3. Select an application domain and show how the appropriate heuristics would apply to that domain.

Chapter **9**

Governance

> Resilience cannot be engineered simply by introducing more procedures, safeguards and barriers. Resilience engineering instead requires a continuous monitoring of system performance, of how things are done.
>
> *David M. Woods (2006a, p. 348)*

In Chapter 8, Resilience Architecting, we discussed the *organizational decision-making* heuristic, that is to say, that organizational decision making should be monitored. This heuristic is based on the almost universally accepted idea that human decisions can be flawed.

In Chapter 1, On Resilience, we noted the paradox of humans in the system. On one hand, humans are often a necessary component of a system because they have the capacity to analyze and react to totally unexpected disruptions. On the other hand, they can be the source of the disruption themselves.

So why do humans make flawed decisions? First of all, humans are not machines. They cannot be programmed with a certain outcome. Second, there are the cultural factors, such as those described in Chapter 5, Culture.

With all of these factors in mind, organizational planners have put mechanisms into place to mitigate the effects of human errors. These include training and review processes that will be covered in this chapter.

Governance refers to the review and oversight of all activities in any phase of a system's life cycle. During the development phase of technological systems, governance is common in the form of design reviews, peer reviews, and independent reviews. In other domains, governance may take completely different forms, for example, supervisory oversight and other forms. The question is: What should be reviewed, and how detailed should the reviews be? For robust resilience, increased detail may be required.

Architecting Resilient Systems: Accident Avoidance and Survival and Recovery from Disruptions.
By Scott Jackson

Although some ideas in this chapter may seem to apply mainly to technological systems or to the infrastructure that builds these systems, the principles apply to any domain. The idea of cross checks, or detailed reviews, in hospitals, for example, is discussed. Independent reviews, in particular, could apply to any domain.

9.1 INSPECTING IN QUALITY

There is a saying that "you cannot inspect in quality." The notion behind this saying is noble, albeit somewhat idealistic. The notion is that if the design process is really high quality, then the design would be perfect the first time and the need to inspect the product would be completely unnecessary. The logic could be extended to conclude also that design reviews and testing would also be unnecessary.

But then, reality sets in, and one realizes that design processes are imperfect and humans are imperfect. All the paradigms listed in Chapter 5, Culture, come into play to make designs that are imperfect and vulnerable to the catastrophic failures that are the focus of system resilience.

However, the basic truth of the saying has to be recognized, namely, that if you rely entirely on inspections, reviews, and testing, the system will be unreasonably costly and perhaps still vulnerable to catastrophic failure. Hence, one cannot rely entirely on one leg of the system resilience triangle: capabilities, culture, or infrastructure. They reinforce each other. This chapter will show how governance, at different levels, contributes to system resilience.

This chapter will discuss two major topics, first, the idea of independent reviews, what they are, and how they contribute to system resilience. The second topic has to do with the level of detail of reviews and how these reviews uncover design defects.

9.2 INDEPENDENT REVIEWS

Although the concept of independent reviews is one of the basic system resilience capabilities discussed in Chapter 6, Capabilities, this chapter provides more detail and rationale into why they are so critical to system resilience.

The first premise is that individuals rarely have the objectivity always to make completely rational decisions. The first obstacle to rational decisions is the cultural environment itself. Chapter 5, Culture, describes many cultural paradigms that are often detrimental to system resilience. One of the most obvious is the risk denial paradigm, which stands in the way of risk management. As described in Chapter 5, this paradigm is the belief, however misguided, that successful programs do not have risks and that to admit to the existence of risks is a sign of failure. Vaughn (1996, p. 82) points out how risk acceptance was considered to be "normative" on the Challenger program. Hence, a mechanism is needed to add an element of objectivity to the risk analysis.

The second factor affecting objectivity is simple program pressures, such as cost and schedule. In situations in which funds are short and schedules are tight, programs tend to take shortcuts. This is not to say that the persons mandating these short cuts were making consciously risky decisions; they may have believed that the shortcut was justified. Only a second set of eyes can determine whether the decision was correct.

Another factor might be called the limited-visibility factor. Because of limited expertise or experience, a program might not be aware of all the aspects that may affect a program that an outside person would notice. This factor is particularly important when an area of expertise outside that of the program team may affect the system. Two notable examples come to mind: The first is the Tacoma Narrows bridge disaster of 1940 described in Chapter 10, Case Histories. In this case the civil engineers who designed the bridge were apparently unaware of the subtle and complex aerodynamic effects that were at work there. The presence of an experienced aerodynamicist might have changed the outcome. A second example is the Comet aircraft also described in Chapter 4. In this case, the presence of a metal fatigue specialist might have saved the day.

So what is an independent review? An independent review is an inspection by outside persons of all necessary aspects of a program, either technical or nontechnical, that may affect the success of the program. Safety experts would certainly have a primary function on the independent review team.

The basic requirement for a reviewer is that he or she should be completely independent of the program. The most common method is to have a team of experts that work for the same company but not on the same program. Customer personnel may also act as independent reviewers.

Normal independent reviews are held on a predetermined schedule. Performing reviews immediately before design reviews is typical. Review data should be provided to the reviewers in advance, preferably up to 4 weeks in advance. This method gives the reviewers time to study the material and to ask the design teams to address the subject before the actual reviews. Reviewers should not attend review meetings with the intent of raising new questions. These questions should have been resolved before the review.

Special independent reviews may be held under a variety of circumstances. The first circumstance is when management fears that the program is in trouble and needs help. The manager then requests a review. The second circumstance is when the program itself recognizes that there is a problem that can only be resolved through outside intervention. The program then receives a review. However, the primary bodies responsible for calling for independent reviews are infrastructure nodes responsible for resilience discussed in Chapter 7, Infrastructure. Each of these teams maintains a broad view of problems on a program and is therefore most qualified to call for independent reviews.

Another special-purpose review is, for example, to review all risks on a program. If the program has a risk committee, then the independent review team would review all material addressed by the risk committee.

Another important function of the independent review team is to review the findings of incident investigations. These investigations invariably result from incidents especially in operations either catastrophic or minor. The task of the team is to determine whether the cause was symptomatic or systemic, as described by Leveson (2002). If the cause is systemic, then the corrective action may be more extensive than if the cause had been merely symptomatic.

So what authority would an independent review team have? In matter for which a critical safety issue is involved, the independent review team should have the final word. For example, the team would have no-launch authority for missions with human crews. At a minimum, the team would have unfettered access to all technical data. Feynman (1988), for example, provides a detailed account of his difficulties in gaining access to data at the National Aeronautics and Space Administration (NASA) during the Challenger inquiry. This aspect could be difficult for programs involved in classified military activities. However, this aspect should be known in advance and planned for.

Finally, the team should have access to program management so that their recommendations are presented to the highest level of management without any dilution or reinterpretation of the results.

9.3 GOVERNANCE AND CULTURE

As discussed in Chapter 8, Resilience Architecting, there are generally two views of how negative cultural aspects can be approached. The culture can either be changed or it can be contained. Although cultural change is the preferred route, it is far more difficult than cultural containment. Hence, governance emerges as the primary cultural containment method. It is viewed that if a culture cannot be changed, then the next best alternative is to make sure that the negative effects do not have a detrimental effect on the system. The processes described in this chapter are intended accomplish that goal.

9.4 A THREE-TIERED VIEW OF TECHNICAL REVIEWS

Chapter 6, Capabilities, identifies technical management as one of the key capabilities necessary for system resilience. One primary responsibility of technical management is technical reviews. Within systems engineering, technical reviews represent gates between phases of a program at which the maturity of the design is assessed. They also represent milestones at which requirements and design solutions are flowed down through the various levels of the system architecture. Hence, each design review addresses increasingly detailed levels of design.

At the highest level are the design reviews. Figure 9.1 calls these system-level reviews in the sense of the complete system, that is, the aircraft, space system, or nuclear power or chemical plant. Many sources, for example, Jackson (1997), give summaries of the orthodox design reviews.

Level of Detail			
	System Level	System Level Parameters	SRR, SDR, PDR, etc.
	Design Level	Specifications, Drawings	Peer Reviews
	Detail Level	Calculations, Planning, Production Procedures	One-on-one cross check, Inspections
	Level	Parameters and Documents	Types of Reviews

Figure 9.1. Defect review levels.

9.4.1 System-Level Review

The key point about system-level design reviews is that they are the highest level type of reviews and have a built-in level of detail to which they apply. Programs use them as a communications medium with their customers to gain concurrence on the general level of design maturity, to review actual design concepts, and to gain agreement on requirements, risks, and other subjects. Detailed technical data are normally not discussed in these reviews but are presumed to take place outside the reviews. If a subject of critical interest emerges from the detailed work, then it will be covered in the review.

Ideally, customers and other outside people will receive the system-level design review information 2 to 4 weeks before the actual event. In practice, schedule pressures often prevent this from happening so that critical items will have to be discussed in the meeting itself.

In short, system-level design reviews represent only one level of review necessary for system resilience. As discussed previously in this chapter, the independent review capability plays a major part in design reviews. For system resilience, independent reviews would be a standard feature of design reviews. The key point to remember is that catastrophic failures rarely happen at the system level—even though these reviews are necessary—so that reviewing at increasingly detailed levels is absolutely essential.

9.4.2 Design-Level Review

The second level of review is called the design-level as review, as shown in Figure 9.1. Design-level reviews are more detailed than the system-level reviews. They would review, for example, drawings, computer codes, and specifications. The type of review normally used at the design level is the peer review. By definition, peer reviews involve the participation by experts from many different fields. Although independent reviewers from other programs are not automatically part of the peer review, they would certainly be called on if deemed necessary by the resilience focal (see Chapter 7, Infrastructure) or another person of authority.

The idea behind a peer review is that any one piece of information may require the inspection and approval of many disciplines. As a simple example, if an electrical conduit is attached to a piece of structure, then, as a minimum, this connection would have to be reviewed by the electrical engineers and the structural engineers. But, safety issues and human factors issues may be associated with this connection and may require the review of those disciplines as well.

Rather than having presentations, as in the case of the system-level reviews, peer reviews are normally handled with simple inputs from each discipline on a common database. The results are agreed on and then finalized.

A simple rule of peer reviews is that silence is consent. If any discipline fails to make a comment, then its concurrence is presumed.

The peer review is one way to approach the difficult subject of unanticipated and undesirable interactions between components. In the case of the Mars Polar Lander, for example, as discussed in Chapter 4, Case Histories, a peer review between the structural engineers and the software engineers is likely to have identified the interaction between the strut and the software that resulted in the failure of that mission.

9.4.3 Detailed Review

The third level of review is called a detailed review, as shown in Figure 9.1. Rather than reviews, these might better be called cross checks. Once again, this is the level at which faults most often responsible for catastrophic failures often occur. In most high-consequence industries, these sorts of cross checks already exist. In the nuclear power industry, for example, all calculations must be checked by a person of equal or greater expertise and signed by that person. Typical calculations might include, for example, structural, aerodynamic, hydraulic, and other such parameters.

The presumption behind these cross checks is that errors at a detail level may be critical but not visible at a higher level of review, such as the peer review mentioned above. There is a story, perhaps apocryphal, about the calculation for the width of a beam on a bridge. The correct width was 14 inches. However, because of an error in drawing, a line was drawn through the digit "1" in the number 14 leaving only the solitary 4. The result was, according to the story, that the beam was built 4 inches wide rather than 14 inches wide, and it failed.

Now, an astute engineer might have caught this error in the peer review. However, the peer review does not normally cover the calculations, so the thin beam passed the peer review. However, with a good calculation cross check, this error was more likely to have been caught. Although this story may or may not be true, it is indeed the sort of error that sometimes results in catastrophic failure.

Patterson et al. (2007) show the importance of cross checking in the health care domain. In the process they call collaborative cross checking, pairs of individuals with different perspectives examine the others' assumptions or actions to assess validity or accuracy. The notable aspect of this example is that it had nothing to do with the quality of a technological product, but rather a medical process in which the error rate is high.

One method of detailed review suggested in Chapter 3, Disruptions, is the use of the multiple-agent N^2 diagram. The idea here is that all potential and *predictable* desirable and undesirable interactions among components can be identified if this method is applied rigorously. Chapter 3 discusses some of the difficulties and approaches for performing this analysis. This method is a way to implement the *avoid hidden interactions* heuristic discussed in Chapter 8, Resilience Architecting.

In the context of system resilience, let us not forget that reviews at any level go far beyond the design of the end product. Detail errors can occur at any point in the system process. Take, for example, the miscommunication between the night crew and the day crew in the Piper Alpha disaster discussed in Chapter 4, Case Histories. This is an example of a process to assure detailed communications. This type of cross check is equally important as the design of the off-shore platform itself.

9.5 FOREIGN OBJECT DEBRIS (FOD)

Relative to the discussion of detailed reviews (or cross-checks) discussed above, the term foreign object debris[1] (FOD) is widely used in the aircraft development and operations domains. FOD refers to small objects, such as screws or paperclips, which may be ingested into the engines of aircraft and cause severe damage. FOD is particularly important on aircraft carriers. Weick and Sutcliffe (2001) note the attention to detail on aircraft carriers. One of their themes is that if all organizations were run with the same attention to detail as aircraft carriers or nuclear submarines, there would be far fewer accidents.

The idea of FOD can also be applied to the defects that may occur at the detail level in Figure 9.1. These defects are generally human errors that lie outside the best-intentioned processes to define and verify product requirements. One example is basic engineering calculations. If errors in calculations are large enough or obvious enough, they could be uncovered in design reviews or peer reviews. However, for small, but perhaps important, calculations these errors may go undetected. The nuclear power industry, for example, requires that all calculations be double checked and signed off by another person of equal or greater expertise

[1]FOD is also ometimes defined to mean foreign object damage to decribed the consequences of the debris

than the one who made the original calculation. In production, all manufactured items are supposed to be checked against the original drawings by an inspector. If the manufacturer deems the mechanic to be experienced and trustworthy, then some inspections may be eliminated. But what if a new mechanic comes on the job? Are the inspections resumed? Perhaps so, perhaps not. All of these scenarios are potential areas for defects, perhaps catastrophic defects.

The point is that the types of defects discussed above will most likely never show up in a reliability analysis. Reliability analyses normally assume that systems are built as designed, well maintained, and well operated. If any of these is not true, then the likelihood of failure will increase. This is the idea that Richard Feynman was trying to express when managers said that the probability of a space shuttle failure was 1 in 100,000. Engineers, on the other hand, estimated this probability at about 1 in 100. History has shown that the engineers were closer to the truth.

9.6 RESPONSIBILITIES FOR GOVERNANCE

The responsibility for governance will vary greatly by domain.

Although Chapter 7, Infrastructure, discusses organizational responsibilities, it is wise to review the responsibilities for governance.

9.6.1 Resilience Node

There does not actually have to be an official node of the infrastructure called the resilience node or anything else. This is just a designation for the cooperative effort that is required among all nodes of the infrastructure. Whether there is an actual team or not is not important. In this role, the node is responsible for collecting system resilience metrics from all nodes and for addressing corrective actions, especially if the responsibilities should cross organizational boundaries. Hence, independent reviews may cover development, operations, and support.

For human-intensive systems, such as hospitals and emergency infrastructures, this collaboration should exist outside any organization because the infrastructure crosses organizational lines as described in Chapter 7, Infrastructure. Hence, the infrastructure is a system of systems.

One main responsibility of the resilience node is to call for independent reviews for issues that may affect any node of the infrastructure. This team may then call for actions to correct the defect that resulted in the independent reviews.

9.6.2 Program System Resilience Team

This team performs basically the same function as the joint resilience node but at the program level. One difference is that the program-level team has direct responsibility for the functional teams on the program as well as the suppliers that provide subsystems and components to the program.

This team will also call for independent reviews and actions to correct flaws, especially if they are systemic in nature. One primary function of this team is the definition and collection of metrics, as described in Chapter 10, Measuring Resilience. This team will use these metrics as a way to identify program changes, that is, ways in which the program is not adhering to the principles of system resilience, that is, the principles outlined in this book.

This team will respond to the joint resilience node for the resolution of any issues affecting system resilience.

9.6.3 Program Management

Program management has the toughest job of all: It has to embrace and implement the findings of the reviews. Sometimes, these require financial and schedule decisions, which may be difficult. Other important functions of program management are as follows:

- Authorize the system resilience organizational structure described in Chapter 7, Infrastructure.
- Authorize the cultural initiatives described in Chapter 5, Culture.
- Authorize the capabilities listed in Chapter 6, Capabilities.
- Authorize the chartering of the joint system resilience team and the program system resilience team discussed in Chapter 7, Infrastructure, and the open flow of information among them.

9.6.4 Functional Groups and Integrated Product Teams (IPTs)

In the civil infrastructure domain, this category would apply to, for example, a battalion of firefighters or a police squad.

- The primary function of these groups and teams is to own and execute the capabilities listed in Chapter 6, Capabilities.
- They will also conduct peer reviews and design-level cross checks as described above in this chapter.
- They will provide experts to participate in independent reviews, providing the reviews are for a program other than the one the groups report to.
- They will respond to actions from the program system review team.

9.7 FURTHER EXPLORATION

After you review the case studies in this book as well as others you have uncovered in your research, identify the kinds of reviews that might have been especially helpful in those cases. Explain why.

Chapter **10**

Measuring Resilience

We can only measure the potential for resilience but not resilience itself.

—David M. Woods (2006a, p. 348)

One of the most common questions is: How does one measure resilience? The tacit assumption behind this question is that system resilience is something like reliability, that is to say, the probability of system failure can be quantified with a number like 10^{-9}. This issue came to the fore with Richard Feynman's (1988) estimate that the chances of a Shuttle failure were closer to the National Aeronautics and Space Administration (NASA) engineers' estimate of 1 in 100 compared with management estimates as low as 1 in 100,000. The root of the difference in these estimates is a cultural one. We saw in Chapter 5, Culture, that a common mindset is to view failure in purely technical terms, for example, reliability. So with two failures in about 100 launches, the engineers' estimate was much closer. Evidently, Feynman had a good feel, albeit intuitive, about the human factors involved in catastrophes. As we saw in Chapter 5, these factors can include poor risk management, deficient communications, misguided decisions, and many others.

10.1 WHY MEASURING RESILIENCE IS DIFFICULT

Measuring resilience is difficult for many reasons, we will summarize here. This is not to say that there are not many good ways to measure, at least qualitatively, the resilience of a system. In addition, certain aspects of resilience *are* measurable. This chapter will discuss those also.

Architecting Resilient Systems: Accident Avoidance and Survival and Recovery from Disruptions.
By Scott Jackson

10.1.1 The Measurement of Human Aspects is Limited

According to Flin (2007), human factors are responsible for 70% to 80% of accidents, and technical factors are responsible for 20% to 30%. Among the human factors, about 80% can be attributed to organizational and cultural factors, and about 20% are operator errors.

We also saw in Chapter 4, Case Histories, that reliability failures can sometimes, but not always, be the source of a disruption. At the same time, we saw that good reliability does not ensure resilience. As a matter of fact, most failures occurred as a result of human intervention or negligence at various stages in the life cycle of the system: development, production, maintenance, and operation. In these cases, humans were rarely observed to be the root causes but rather were factors in larger systemic causes.

We should say once again that reliability cannot be neglected. To be resilient, a system should first be reliable. That is why reliability is listed as one of the necessary analytic methods discussed in Chapter 6, Capabilities. However, reliability is not sufficient. Many other human-centered factors need to be treated to make a system resilient. For human-intensive systems, the human aspects are the central consideration.

Because human aspects are absent from reliability analyses and to a large extent from safety analyses, both these types of analyses will tend to give optimistic assessments of the likelihood of success of any system. Feynman's observation discussed above confirms this conclusion.

The conflict between quantitative resilience and qualitative, human-focused, resilience is the issue addressed by the *design-focus paradigm* discussed in Chapter 5, Culture. Thus, specialists who are design-focused may have difficulty with qualitative results.

In their "initial introduction" to measuring resilience, Woods et al. (2006, p. 347) address the difficulties by identifying parameters within an organization, e.g., number of personnel, procedures, etc., and suggest that failure points can be predicted in a manner to stress-strain plots in materials.

10.1.2 Heuristics do not Always Lend Themselves to Quantitative Analysis

We saw before in Chapter 6, Capabilities, that the discipline of systems architecting utilizes *heuristics* as part of the decision-making process. Heuristics are, by definition, not quantitative. They are simply lessons from the past.

Chapter 8, Resilience Architecting, provided a set of heuristics that has been derived from the literature for architecting resilient systems. Typical of these is that the system should be able to reorganize itself. Reorganization is not an easily quantifiable attribute, if at all. However, experience has shown that it is valuable in the face of disruptions.

In short, although it is agreed that these heuristics will make a system more resilient, a quantitative measure is not easily obtained.

10.1.3 Unpredicted Disruptions Lead to Unpredictable Outcomes

We saw in Chapter 3, Disruptions, that failures are initiated by disruptions. And sometimes, disruptions can be completely unpredicted.Chapter 3 provided examples of disruptions that were caused by humans, software, and hardware. Hence, it only follows by simple logic that if characteristics are not predicted, the probability of their occurrence cannot be quantified.

To put this idea into mathematical terms, assume that there is a relation $y = f(x)$, where x is the magnitude of a disruption and y is the magnitude of the result, that is the stress on the system. If the value of x is unknown or unpredicted, then the result will also be unknown.

So, is there value in designing a system for unpredicted disruptions? The answer is "yes" because the heuristics tell us so. So, how much do the heuristics help us? As we saw above, that is difficult to say, but it will be better.

10.2 APPROACHES TO MEASURING RESILIENCE

Despite these difficulties, there is no shortage of authors who have suggested many useful approaches.

So, how do we arrive at a quantifiable measure? What is the probability that schedule pressures will result in a deleted critical test, for example? What is the probability that a complex interaction between two components will escape the engineer's attention? There are two answers to these questions and others like them, and they all are more complicated than a reliability analysis. The first is by statistical prediction. The second is by risk analysis. Both Paté-Cornell (1990) and Leveson (2002) show us their own versions of risk analysis.

This book does not suggest that any particular method of measuring system resilience is a panacea. All these methods are to some degree either in the formative stages or are purely notional. Another limitation of quantitative methods is that the inputs are often highly subjective. Despite all these limitations, the pursuit of methods to measure system resilience should remain a high-priority task. Otherwise,the question "have we done enough?" will remain unanswered.

10.2.1 Probabilistic Risk Assessment (PRA)

Paté-Cornell (1990) is an advocate of probabilistic risk assessment (PRA). While using the equations of conventional probability analysis, Paté-Cornell assigns probabilities to the detection of design errors by various people in the design review process. However, the people involved were the lead engineer, the engineering manager, and the constructor. She divides errors into two categories: gross errors and errors in judgment. Finally, she categorizes the errors into high severity and low severity. Using these values, she can calculate the probability of the failure of an entire system, in her case, off-shore platforms. This method could be expanded to include errors by various other people in the process, for example, operators and maintainers.

Paté-Cornell (1990) notes that a methodology of assigning probabilities of errors in judgment to managers is not likely to gain acceptance in industry. This is an example of where a cultural factor (see Chapter 5, Culture) comes face to face with a science-based methodology. The increased use of independent reviews is one way to negate this effect as noted in Chapter 9, Governance.

Paté-Cornell (1990) also notes that it is impossible to assign a probability if a failure mode has not been identified. This fact reinforces the need for a rigorous system safety capability as noted in Chapter 6, Capabilities.

Finally, Paté-Cornell is the first to point to the "softness" of this methodology. That is to say the values lack credibility and "can be manipulated." Assigning probabilities to human error can be particularly difficult. She does say, however, that there is some evidence to support some of the data.

Paté-Cornell and Fischbeck (1994) use the PRA methodology to analyze the tiles of the Columbia Space Shuttle vehicle. They concluded that changes in the maintenance procedure could reduce the probability of failure by about 70%.

All the above having been said, what is the value of such methodologies? In short, it does support the basic finding, one of the themes of this book, that system resilience is not a purely technical characteristic and that factors such as organizational capabilities, culture and infrastructure, do have a major, if not the major, effect on it. Any efforts, such as this one, to quantify resilience are therefore of value.

10.2.2 The Massachusetts Institute of Technology (MIT) Risk Model

Leveson (2004b) describes a methodology for assessing the risk of systems to major catastrophes based on advanced logic. Hence, the implied measure of system resilience is risk; the lower the better. The methodology is embodied in a simulation called Systems-Theoretic Accident Modeling and Processes (STAMP). According to Leveson (p. 1), traditional methodologies focus on "failure events in static engineering designs and linear notions of causality"; that is to say, traditional methodologies normally assume that all probabilities are independent.

According to Leveson (2004b, p. 1), new tools and models "need to treat systems as dynamic processes that are continually adapting to achieve their ends and to react to changes in themselves and their environment." STAMP is such a tool. STAMP describes "the process leading up to an accident as an adaptive feedback function that fails to maintain safety constraints as performance changes over time to meet a complex set of goals and values."

Leveson (2004b, p. 5) applies STAMP to an analysis called STAMP-based Analysis (STPA), which "enables model-based simulation and analysis of risk throughout the system life cycle, including complex human decision-making, software errors, system accidents (versus component failure accidents), and organizational risk factors." Hence, using STAMP, STPA goes far beyond traditional technical risks to include human aspects, such as decisionmaking and other organizational risk factors. Hence, STAMP requires that these human and organizational aspects be characterized for inclusion in STAMP. To accomplish this characterization, Leveson et al. (2006) say that STAMP uses visualization and shared mental models of those responsible for managing risk.

Central to risk management is the concept of control. Leveson et al. (2006) say that accidents, such as Challenger, occur from inadequate control of the development process; that is, risk is not adequately managed in design, implementation, and manufacturing. The purpose of STAMP is to characterize this lack of control.

Leveson et al. (2005) provide a description of the risk model within STAMP. Following are some inputs to the model:

- Inspection quality (morale, customer and contractor expertise, complacency, oversight capacity, and resources)
- System safety efforts and efficacy in isolation
- Effect of symptomatic action on risk
- Effect of systemic corrective action on risk
- Time for changes to affect risk
- Effect of system age on risk

It is easy to visualize the wide range of parameters that need to be defined and entered into the model. The net result is a level of risk that is time dependent and that shows the interaction among various factors as described above.

10.2.3 Statistical Prediction and the Iceberg Theory

A topic among resilience researchers is whether resilience, or better put, a lack of resilience, can be predicted.One school of thought expressed at Juan-les-Pins in 2006 was that accidents are so random that any type of prediction is out of the question. However, others point to defects and near misses that almost always precede major accidents. This idea is called the iceberg theory, a term used by Van der Schaaf at Juan-les-Pins. Can it be said, for example, that an aircraft that requires more maintenance is more likely to have an accident? The answer to this question is not known, but it is possible. There is a smattering of actual statistical evidence, for example, the Scottish Railways study, which indicates that there is some statistical correlation between minor events and major accidents. Figure 10.1 is based on the results of Wright and Van der Schaaf (2004). What this figure shows is that the incidence of minor defect, called near misses, correlates very closely with major accidents resulting in fatalities and injuries for the same causal codes. The correlation is very strong for all levels of data, which include technical, proximal, intermediate, and distal. The latter categories have to do with the breadth of the information. Even with this limited data, it seems clear that the iceberg theory does have some validity.

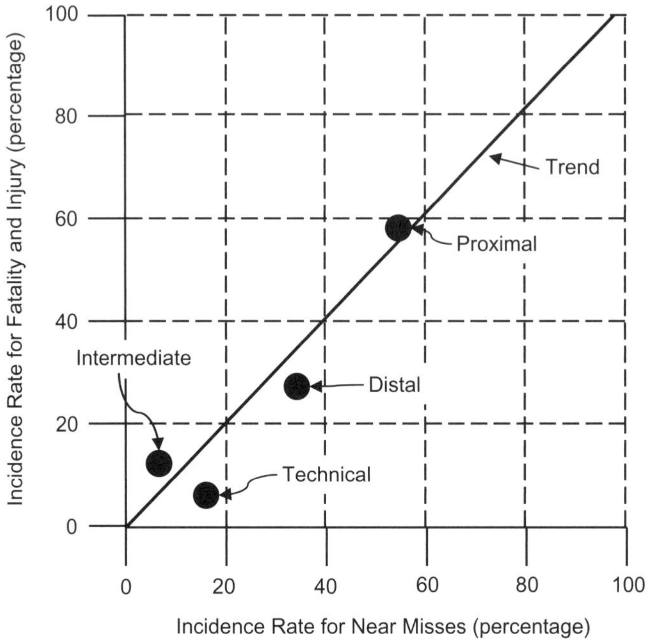

Figure 10.1. Causal code incidence rate correlation. Adapted from Wright and Van der Schaaf (2004).

If developed, such evidence would have enormous benefit. This information could be used to create better designs, better operational procedures, or better maintenance procedures. In short, it could be used to produce more resilient systems.

The basis for statistical prediction is a familiar concept. First, Reason (1997, pp. 118–119) states that "because [near misses] occur more frequently than bad outcomes, they yield the numbers required for more penetrating quantitative analyses." Reason uses the term *near misses* to refer to all defects and incidents, however small, which may be indicators of larger problems. Reason defines near miss as "any event that could have had bad consequences, but did not."

In addition, Leveson (1995, p. 242) says "Examination of and understanding of near misses can warn of an impending accident. . ." The air traffic domain probably uses the term "near misses" more often than any other. Wald (2005) says that the Federal Aviation Administration (FAA) reports that the number of aircraft flying too close to each other in the New York area was more than six times as high as previously reported. Although the number of midair accidents had not increased, the FAA does consider near misses as key measures of resilience deficiency.

So, we are starting out with the promise of a method that is, as yet, not extensive validated. The proposed method starts with the premise that little problems (near misses) are indicators of big problems (catastrophes). Leveson et al. (2006, p. 121) lends credibility to this premise. In her risk model, she shows that when "systemic factors" are fixed, the risk is much lower than when only symptoms are fixed. Systemic factors do not have to be actual system failures. They can be, for example, inadequate design reviews or inadequate inspections. Leveson also points out that getting agreement on what these factors are can be difficult. Engineers, pilots, and managers will all have different ideas about what these factors are. Once again, independent reviews will be of assistance in coming to an agreement on these factors. See Chapter 9, Governance.

A dramatic example of symptoms versus systemic factors is the Mars Climate Orbiter mission described in Chapter 4, Case Histories. The symptom was the transmission of flight control data by English rather than metric units, whereas the systemic cause was to have a robust interface management process in place.

Another aspect of selecting low-consequence metrics is whether they are predictive or reactive. Predictive metrics are those that are collected prior to the system's being introduced into the operational phase. They could have to do with the design reviews or inspections. For example, reactive metrics have to do with failures in operation, such as breaks in a fuel line, as mentioned above. Reactive metrics are, of course, also predictive in the sense that they may predict future operational failures. If a system is in the conceptual phase, then all the metrics will be predictive. If the system is in the operational phase, there will be both predictive and reactive metrics.

The following are a few examples of defects or near misses that might have been measures of resilience. The classification into technical, proximal, intermediate, or distal is, for the purposes of this book, judgmental and not authoritative, but it serves to illustrate the concept. These defects may be process defects, management defects, operational defects, or design defects. Although most catastrophes result from a combination of defects, only one will be listed here for illustration.

Piper Alpha, the North Sea Oil Disaster. According to Paté-Cornell (1990), a key defect in this disaster was a lack of communication between maintenance crews.

Challenger. According to Vaughn (1996), the main defect was a lack of verification of critical components.

Mars Polar Lander. According to Leveson (1995), the defect contributing to this accident was the unpredicted interaction between the struts and the software.

Nagoya. As described by Leveson (2004a, p. 16), the principal contributing defect was the fact that the pilot's mental model did not match the flight-control algorithms.

The defects in the above accidents will be classified as proximal, distal, intermediate, or technical depending on the level of detail of the evidence.

10.2.4 Using Accident Metrics: The Commercial Aircraft Experience

Probably the most dramatic use of accident metrics to identify safety enhancements is documented by the Commercial Aviation Safety Team (CAST) as described by Matthews and Romanowski (2004) who report that the rate of fatal accidents has been reduced by 65% over a ten-year period. For example, one method of reducing accidents is to tailor the approach angle to airports so that aircraft do not exhibit the unstabilized approaches that were associated with a high percentage of accidents. CAST (2007) is an international cooperative initiative with 23 industry and government organizations as members. CAST analyzes accident metrics and makes recommendations for safety enhancements.

In its search for safety indicators, the FAA (1997) has concluded, first, that the fatality rate of individual carriers over time is so unstable that it is not possible to use carrier identification as a an indicator of fatalities. However, it was found that comparing carriers by groups does reveal some differences. For example, U.S.-based airlines generally have lower accident rates.

Interestingly, the FAA report (1997, Section: Recent Research) states that "To date there has been relatively little research into relationships between accidents and less serious safety measures such as incidents or surveillance data." This is to say that limited data are available in the commercial aviation domain similar to the Scottish Railways study discussed above. Nevertheless, the FAA has reached one conclusion that there does not seem to be any statistical correlation between a potential predictor, the near midair collision (NMAC) and actual midair collisions (MACs). More interestingly, the FAA has concluded that the statistical correlation between nonfatal incidents and passenger death rates is negative. It can be concluded that fatal accidents are caused by different phenomena from nonfatal incidents.

The information collected by the FAA, such as inspection and surveillance reports, would be most useful in generating the correlations suggested by the iceberg theory, above. However, these analyses are ongoing and are not available to the public. However, much of these data are collected by the Department of Defense (DoD) and are available. The DoD has concluded that there are five broad metrics with which to score carriers. They are as follows:

Safety
Operations
Maintenance
Financial condition
Service quality

Similarly, according to the FAA safety report (1997) the U.S. General Accountability Office (GAO) has four similar measures, which are as follows:

Pilot competence
Maintenance quality

Financial stability
Management attitude

The striking thing about these two lists is their similarity. However, both of these lists were put together using the judgment of the two agencies involved, not rigorous statistical analysis. Following are some comments on each one.

With regard to safety, the FAA comments that there is a scarcity of safety measures. The concept they suggest is that this metric would be based on FAA inspections and surveillance and would reflect compliance with safety regulations.

Although the DoD uses the term *operations*, the FAA suggests that this term pertains to all the other factors listed. However, it can also be taken to mean data taken during the operational phase of the aircraft life cycle.

The FAA lists pilot competence as a separate factor. This factor can be important in the resilience of a system as it was in the case of U.S. Airways Flight 1549, as discussed in Chapter 4, Case Histories.

Both lists mention maintenance as a separate factor. The inclusion of this factor is in agreement with the cases covered in Chapter 4 in which there were several cases for which maintenance was a contributing factor to an accident. American Flight 191 and the ValuJet cases are examples. Service quality can be recognized as an extension of maintenance, so the same conclusions will apply. Chapter 7, Infrastructure, makes the case that a possible cause of maintenance errors is the inherent weakness of the interfaces between the nodes of the development infrastructure system.

The financial stability metric can be observed in the context of the *conflicting priorities* paradigm discussed in Chapter 5, Culture. The premise of this paradigm is that if there is a conflict between, for example, financial needs and system resilience, then the financial needs will take top priority. If either a private company or a government agency has limited or financial resources, resilience may suffer.

Management attitude is an intuitively correct metric. However, it is not obvious how this metric could be quantified. Many aspects of management attitude are discussed in Chapter 5, Culture. It is assumed that these attitudes could be measured through responses to questionnaires, interviews, and other such techniques.

The common element among all the above metrics is that they are all intuitively useful. However, one strength of statistics is that it can extract effects that may not be intuitive. As the FAA report (1997) points out, there are analyses being conducted that have not yet been made public. It is assumed that rigorous in-depth statistical analyses are among them.

10.2.5 Using Metrics to Manage an Organization

So the next question is: What do you do with the metrics once you have collected them? The answer is that you use them to manage your organization and to improve system resilience. As shown in Figure 2.2, the first purpose of metrics is to measure system resilience in whatever way it can. This chapter provides several methods of doing that. Second, the metrics are used to make improvements in system resilience capabilities.

The beauty of using the discussed metrics methodologies to manage an organization is that it is not necessary to base the metrics or the recommended organizational changes on actual catastrophes. Rather, it is possible to make organizational changes based on just symptoms that may be reliable indicators or predictors of catastrophes. This is especially true if the indicators are predictive, rather than reactive, as discussed above. Leveson et al. (2006, p. 121), for example, using risk analysis shows that "systemic factors" can be found that are good predictors.

So what kinds of organizational improvements will result from metrics? First of all, using metrics to drive organizational changes is not new. It is a very popular way to determine desired organizational process improvements. It is not clear, however, that the organizational changes presently being contemplated are focused on system resilience improvement.

The desired organizational changes depend on three factors: first, the measured systemic factors themselves. That is, which systemic factors have been found to be the best predictors of potential catastrophes? Second, what organizational procedures or process improvement will actually result in an improvement in the systemic factors and, hence, in an improvement in system resilience? Finally, what is the most cost-effective way to allocate funds to achieve the greatest improvement in system resilience? We shall discuss each one of these three factors.

First, how do you identify the systemic factors you are going to analyze with respect to catastrophes? Leveson (2002) points to the difficulties in doing this. She cites studies that show that the identification of the causes of catastrophes is greatly dependent on who is doing the analysis. Engineers and managers, for example, may come to completely opposite conclusions. An important factor in establishing these causes is *independence*; that is to say, only through independent review can really objective views of causes be obtained. See Chapter 9, Governance, for a discussion of independent reviews.

The second factor in determining organizational changes is identifying what exactly has to change to realize an improvement in system resilience. The answer to this question rests in the nature of the systemic factors themselves. In Chapter 4, Case Histories, it was recognized, for example, that communication failures were a major cause of catastrophes. So what has to change to effect an improvement in communications? There are many possibilities, for example:

- Increased use of teams in which discussions across organizational boundaries are conducted.
- Organizational restructuring to assure that the communications paths to decision makers is minimized.

- Auditing of information across key paths within and across organizations. See Chapter 6, Capabilities, for a discussion of information management and how auditing would be conducted.
- Training. Chapter 6, Capabilities also discusses training for improved communications.
- Independent Reviews. Chapter 9, Governance, discusses how independent reviews would improve communications.

The above is just one example of how a root systemic factor can be changed to improve system resilience.

The third and final factor in determining organizational changes is to determine which changes can be achieved in the most cost-effective way. The way to find this out is to use what is sometimes called the "portfolio optimization" method. In short, it is asking the question: For any given amount of money what is the most cost-effective way to reduce risk? Is it better to put $1M into improved testing, or is it better to put $1M into more design reviews? Optimization methods exist that will answer these questions.

10.2.6 Organizational Resilience

Flin (2007) provides a list of indicators of diminishing resilience in an organization. However, these indicators are, to a great extent, subjective and observational. However, if used in a rigorous way, they can be good signals of reduced resilience within an organization. They are as follows:

- Increased primacy of production and cost cutting
- Past success as a reason for confidence
- Limited concern and monitoring of emerging risks
- Suppression and dismissal of warnings
- Fragmented decision-making process
- Problems with change management
- Friction at internal boundaries
- Reduced resources and redundancy

These indicators are clearly subjective and nonquantitative. Reason number 2 on past successes is a clear reference to the Titanic Effect discussed in Chapter 5, Culture. Others are also direct references to other paradigms, such as the risk avoidance paradigm.

The "friction at internal boundaries" item can be related to the brittleness caused by a failure to comply with the *interelement collaboration* heuristic discussed in Chapter 8, Resilience Architecting.

10.3 SOME MEASUREMENTS THAT ARE POSSIBLE

Not surprisingly, the parameters that are measurable are more technical in nature. This fact does not mean they are unimportant.

10.3.1 Reliability

Without a doubt, it is possible to determine the reliability of any system element, especially for technological systems. Although this fact may not be satisfactory to someone concerned about total system resilience, the fact remains that reliability is an essential part of resilience and is measurable. Blanchard and Fabrycky (2006), for example, devote considerable attention to this subject. It is true that components with a significant history of performance have more accurate reliability values. However, new designs are subject to extensive analysis before a number can be established.

10.3.2 Safety

Although safety differs from reliability in that it considers known hazards, it is quantitative and, hence, is useful for resilience.

10.3.3 Capacity

As discussed in Chapter 8, Resilience Architecting, capacity is a basic attribute of resilience. For example, it is possible to determine many features of a system based on historical data. Wind levels in hurricanes and flood levels are examples. These levels can be used to determine structural requirements for buildings and the height of levees.

A typical example of capacity measurement is the analysis of the resilience to floods by the UK Department for Environment, Food and Rural Affairs Defra (2007). The frequency of flood occurrence is well documented and can be used to evaluate the effectiveness of antiflood measures.

So, among all the resilience attributes listed in Chapter 8, Resilience Architecting, capacity is the easiest to measure.

10.3.4 Measuring Resilience in Health Care

The measurement of resilience in health care is decidedly qualitative. Mallak (1998) provides some typical health care metrics. Mallak defines scales for each of the following factors:

1. Goal directed solution seeking—This factor measured the enjoyment of improvising solutions to a problem.
2. Avoidance—This factor measured the ability of a subject to back off from a problem when it becomes too difficult.

3. Critical understanding—This factor measured the ability of a subject to understand the problem when faced with conflicting information.
4. Role dependence—This factor measured the ability of a team member to perform each other's roles.
5. Source reliance—This factor measured the ability of a team member to rely on multiple sources of information.
6. Resource access—This factor measures the ability of a team member to access the information necessary to do the job.

10.4 FURTHER EXPLORATION

In your research of case studies and the defects identified by the sources, identify which defects can be classified as symptoms, systemic, reactive, or predictive.

Chapter 11

Cost

You can meet your performance, cost or schedule goals, but not all three.

—Anonymous

The first question an executive will ask about resilience is how much will it cost? It is not difficult to determine how much a set of new fire trucks costs or a new communications system. But how do you determine how much resilience these trucks and these radios will buy you? This question is so difficult that most authoritative references are only in the initial stages of considering it, for example, Hollnagel et al. (2006) and Hollnagel and Rigaud (2006) pay little attention to it. Other sources, for example, Defra (2008), address only the quantitative aspects of resilience to address the cost of resilience.

The cost of resilience is inextricably tied to the ability to measure it. We saw in Chapter 10, Measuring Resilience, that resilience is difficult to measure. For the very same reasons—human factors, unpredicted disruptions, and heuristic methods—also make cost estimating difficult. Also, for the same reasons, we have those quantitative methods available to use. These methods will not give us the whole story, but they will provide a solid foundation from which to begin.

11.1 LIMITATIONS ON RESILIENCE COST ANALYSIS

We saw in Chapter 10, Measuring Resilience, that certain aspects of resilience are not amenable to quantification. These include unpredicted disruptions, the use of heuristics, and human error. The difficulty of quantifying resilience renders the quantification of cost difficult, at least in these three areas.

Architecting Resilient Systems: Accident Avoidance and Survival and Recovery from Disruptions.
By Scott Jackson

11.2 THE OPTIMISM AND PESSIMISM OF COST ANALYSIS

We have discussed that the traditional methodologies of reliability and safety can provide probabilities of success, at least for technological systems. Human-intensive systems, such as civil infrastructures, have no such foundation in predictability, except for the technological components in them, such as fire trucks and power systems.

So, let us say that a system of any type has been built, and the cost has been determined based on traditional costing methods, that is, not considering unpredicted threats, heuristics, or human error. Do these three factors make the system less resilient or more resilient than previously considered without these factors? Let us examine these factors one by one and evaluate the optimism or pessimism of the original analysis. Because most systems are constructed with these factors either ignored or minimized, this scenario is not uncommon.

11.2.1 Unpredicted Disruptions

First, let us take unpredicted disruptions. In this case, the system is less resilient than originally thought, so the designers were optimistic. Take the case of the twin towers. Obviously, the terrorist attacks were unpredicted, so the system was less resilient than originally thought. The fate of the buildings is the evidence of this conclusion.

11.2.2 Human Error

The second factor is human error. Flin (2007) notes that 70% to 80% of all accidents are caused by human rather than technical errors. So a system designed without consideration of human error is bound to be optimistic, that is, less resilient than it otherwise would be.

It is useful to study the Federal Aviation Administration (FAA) policy on this subject. Following is an excerpt from Federal Aviation Regulation 25.1309 (2007, p. 1):

> The airplane systems and associated components, considered separately and in relation to other systems, must be designed so that—
>
> (1) The occurrence of any failure condition which would prevent the continued safe flight and landing of the airplane is extremely improbable, and
> (2) The occurrence of any other failure condition which would reduce the capability of the airplane or the ability of the crew to cope with adverse operating conditions is improbable.

In summary, this requirement has two points: First, it requires that the likelihood of occurrence of any failure condition should be inversely correlated to the severity of the consequence. Second—and most important—this requirement has to do with the *design*, that is, human error is normally not considered in the calculation of likelihood. However, since the FAA (2007a) requires the probability of an unsafe event to be calculated quantitatively, and human error is difficult to measure, human error is considered qualitatively.

In short, even the FAA normally does not consider human factors to be a factor that can be used in the evaluation of the likelihood of a failure. This is not to say that the FAA ignores human factors. On the contrary, the FAA has extensive programs to incorporate human factors in their policies. However, these programs—for the most part—consider human error in a qualitative way and the system of heuristics to be discussed below.

The final question is whether human error can be or actually is measured. There is little doubt that it can be measured, but it is only measured in specific domains in well-defined situations, such as pilot errors in commercial aircraft and hospitals. These domains will be discussed later in this chapter.

11.2.3 Heuristics

Heuristics, as discussed in Chapter 8, Resilience Architecting, are the design principles that cannot be verified except by experience.

Heuristics are the design principles that have been shown to increase resilience. However, as we have said, this increase cannot be quantified.

So, contrary to human factors and unpredicted threats, a system built without consideration of these heuristics can be said to be less resilient than one for which these heuristics have been employed.

So the quantitative analysis of systems built with heuristics can be said to be pessimistic, that is to say, the system will be more likely to fail than the quantitative analysis would show. Exactly how much more likely the system is to fail cannot be determined.

The only caveat that needs to be added to the conclusion of pessimism is that many systems are built using heuristics today, commercial aircraft and nuclear power plants, to name two. So, to the extent that these designs have already incorporated heuristics, then the failure histories will not be pessimistic, but rather will reflect real failure histories.

11.3 SOME VIABLE APPROACHES TO COST ANALYSIS

Despite the above factors, there have been many efforts to determine the cost of resilience. These efforts have relied almost exclusively on the quantitative aspects of resilience.

The following examples will illustrate how the cost of resilience can be determined.

11.3.1 The FAA Cost Approach

Determining the cost of resilience for technological systems, such as commercial aircraft, is easier than for human systems. The FAA has done this, albeit using solely quantitative data.

The issue arises whenever the FAA has to decide whether to mandate a design change that must be implemented in all commercial aircraft to be certified. Like all federal agencies, the FAA is required by the White House (1993) to show that their regulations are cost-effective. For the FAA the method is very simple. The FAA compares the cost savings with accidents avoided with the cost to implement the change. If the cost savings are much greater than the cost of implementation, they will add or change the regulation or standard. However, there is no known ratio between cost savings and implementation cost that the FAA uses. Typical ratios can be inferred from published reports.

For example, in the analysis of a design change related to protection against icing on the aircraft, the FAA (2007, August 8) found that over a 45-year period there would be a savings of $89.2 million and a cost of rulemaking—that is, implementation—of $62.3 million. Hence, with these results, the FAA implemented a rule change for applications for aircraft of a new type.

A key aspect of determining the cost savings is establishing the value of human life. The methodology for putting a cost value on life was determined by GRA Inc. (2007) for the Department of Transportation. In short, a value of $3 million was used. The cost was based on productive life remaining and medical and other expenses. The value of life numbers are used by the U.S. Department of Transportation for analyses in areas other than commercial aircraft, for example, automobile crashes. This value will undoubtedly change over time.

Chapter 4, Case Histories, provides two cases in which the FAA evaluated the cost effectiveness of mandating a change. One was the Sioux City crash in which the pilot was able to save lives by using the throttle to land the aircraft. In this case, according to National Transportation Safety Board (NTSB) (1990), the NTSB decided that preventive measures were more cost effective. The other case was Trans World Airlines (TWA) 800. In this case the FAA decided to require means to reduce flammable vapors in the fuel tank cavities. The use of nitrogen-enriched air (NEA) in fuel tanks is one such means. However, the specific methods to achieve reduced vapors will depend on cost-benefit analyses.

The FAA mandates design changes if they are required for a safety need; but at the same time they much be cost-effective in accordance with the White House Executive Order 12866 (1993). FAA regulations can mandate changes in two ways. First, there are the general rules that apply to all aircraft with capacities exceeding 30 passengers or 7500 pounds payload. The second type of rules pertains to a specific unsafe condition that may exist on one aircraft and other aircraft of the same type as explained in Government Printing Office

(GPO) (2009, ¶. 39.5) Electronic Code of Federal Regulations. These regulations can be found in the Airworthiness Directives of the Federal Aviation Administration (FAA) (2009). Rule changes resulting from the Sioux City accident did not require cost-benefit analysis.

So is the FAA analysis optimistic or pessimistic? The answer is that it can be considered optimistic from one point of view and pessimistic from another. On the one hand, from the likelihood-of-mishap point of view, it is optimistic because the reliability and safety analyses do not usually consider human error in their probabilities. On the other hand, it is pessimistic because it does not consider the heuristics used in pilot-automation analyses. Nowhere are pilot-machine analyses used more than in the design of commercial aircraft. They are the Billings heuristics discussed in Chapter 8, Resilience Architecting. Because these heuristics are not quantitative, they do not figure into the probability of mishap analyses. As heuristics are always considered to make systems more resilient, the commercial aircraft system will be more resilient than predicted by the numeric analyses. What is the likelihood that the pilot in the Sioux City case would actually save lives? This question is not even implied in the analyses.

So, which factors are stronger, the optimistic factors or the pessimistic factors? The only clue is to examine the record of fatal accidents over many years. Have there been more or fewer accidents than the numeric analyses predicted? The answer to that question is not known.

11.3.2 Cost Based on Quantifiable Measures

The first hurdle to cross on the way to comprehending the cost of system resilience is the idea that system resilience is quantifiable, like reliability. The desire of a design-focus person is to ask "What is the probability of system success?" When that is known, a number can be put on it, for example, one failure in one million missions. The cost of system resilience will then be the cost of achieving that number.

So, the first step in understanding the limitations on costing system resilience is to understand the limitations on quantifying system resilience itself.

As discussed in Chapter 10, Measuring Resilience, there are three basic quantifiable measurements for resilience. If it is desired to base the cost of resilience *only* on these factors, that can be done. In some domains, that may be all that is necessary. However, it should always be kept in mind that the factors of unpredicted threats, human error, and heuristics are not included.

11.3.2.1 Reliability. This term, discussed in Chapter 6, Capabilities, applies primarily to hardware technological systems. It is not meaningful to talk about the reliability of human components. The most accurate reliability numbers come from actual component histories in an operational environment. Although the reliability of a new component can be estimated by analysis, this method is not as accurate as actual field data.

If the reliability of a system does not meet a desired goal, the reliability can be improved through physical redundancy. Physical redundancy is one of the basic heuristics discussed in Chapter 8, Resilience Architecting. Physical redundancy is normally only employed if the system elements are either lightweight, inexpensive, or both.

In short, the cost of achieving a given level of reliability is generally straightforward.

However, even in reliability analysis, operational data are collected under the heading "system failures," that is the operational failures that cannot be attributed to the failure of individual components, but rather to the system as a whole.

11.3.2.2 Probability of Mishap. The concept of the probability of mishap is a standard part of system safety engineering. The reason this parameter is quantifiable is because it is based on a known history of disruptions. Hence, unpredicted disruptions are not considered in the calculation of the probability of mishap.

Similar to reliability, the probability of mishaps applies primarily to technological systems, such as commercial aircraft. We discussed above that the probability of mishaps was one of the primary inputs to the FAA cost methodology for deciding whether to mandate a design change in commercial aircraft.

We saw also that human error is sometimes but not usually considered in the calculation of probability of mishap, at least not by the FAA. However, wherever data exist, it is indeed possible to include it in the analysis and hence use it as part of a cost analysis.

Although the concept of probability of mishap is known primarily in technological systems, and human error is normally omitted in the analysis of these types of systems, there is no reason that it would not be applicable to human-intensive systems, such as hospitals where human error is critical. The hospital domain will be discussed below.

11.3.2.3 Human Error in Hospitals. Hospitals are one environment in which human error has been studied. For example, Okada (2003) first determines the performance-shaping factors (PSFs). These factors most strongly contribute to human error. The factors identified were work environments, operations manuals, design of instruments and interfaces, physical stressors, psychological stressors, skills of operators, operators' character, and others. Okada then performs a mathematical analysis to generate a numerical value for a parameter called human error potential (HEP). Using the HEPs as a performance metric, it is then possible to determine actions to reduce human error and assign a cost to each action.

11.3.2.4 Capacity. Capacity is probably the most straightforward resilience attribute to which a cost can be assigned. Capacity is highly domain specific. In

Figure 11.1. Thames flood barriers.
Source: Department for Environment, Food and Rural Affairs (Defra). (2008, June). *Developing the evidence base for flood resilience: Technical Summary FD 2607.* Retrieved on March 27, 2009, from http://sciencesearch.defra.gov.uk/Document.aspx?Document=FD2607_7321_TSM.pdf.

many domains capacity can be determined by historical data. For example, the historical data on floods in a specific river are a matter of public record. Hence, it is possible to determine the height of levees required to meet the worst-case flood condition. It is then also possible to determine the cost of these levees.

For example, Defra (2008) has calculated the cost of resilience to flooding in the United Kingdom based on historical data. Figure 11.1 shows the barriers built on the Thames to protect against predicted levels of floods. The Defra study also addresses the costs of protection at the property level, that is to say, for individual houses and businesses.

Similarly, from historical data on diseases, accidents, and other demands on hospitals, it is possible to determine the number of personnel and equipment needed at each hospital. These are capacity measures, and the cost of these measures can be determined.

However, calculating the cost of capacity for unpredicted disruptions or for unpredicted levels of disruptions makes costing more difficult and uncertain. For example, the Defra report cites the uncertainty in flood levels that result from potential climate changes. Similarly, what is the capacity cost of future earthquakes in California? Could modern buildings withstand an earthquake of Richter scale level 8.6? It is widely believed so, but that will never be known until an earthquake of this magnitude actually happens.

The most difficult question is as follows: What are the options when the cost of capacity exceeds even the most optimistic of financial assets? For example, there is a great deal of uncertainty regarding the magnitude of possible pandemic, such as of avian flu. Even a modest level of outbreak may exceed considerably the ability of hospital emergency rooms to handle the patient load. If we assume that no new emergency rooms are constructed in the foreseeable future, it will be necessary to fall back on other resilience attributes.

One of these is the *graceful degradation* afforded by the human components. Anders et al. (2006) show that hospital emergency departments have a great deal of natural resilience because of the human-intensive nature of the system.

Outside the hospital, the next best line of defense is to depend on remote medical resources. This method depends on the attribute of intercomponent collaboration. This method requires that a system be set up in advance so that other medical resources will be available quickly. One possibility is a central command and control system to make sure the appropriate people are available and notified. A communications system that is not subject to single-point failures would facilitate this system. In short, there are other ways to alleviate the crisis, but advance planning would be required. In setting up these methods, it would be wise to refer to the resilience attributes discussed in Chapter 1, On Resilience.

11.4 SOME LOW-COST APPROACHES TO RESILIENCE

There is almost no domain that has unlimited funds for development, deployment, staffing, and sustainment. Civil infrastructure elements, in particular, may have difficulty being fully funded. In the space domain also, Leary (2008), for example, describes efforts within the National Aeronautics and Space Administration (NASA) to contain costs.

It is not within the scope of this book to suggest how a specific domain might achieve a greater degree of funding. This book is about architecting resilient systems. However, it is possible to examine certain resilience heuristics in Chapter 8, Resilience Architecting, and arrive at a qualitative judgment about which ones are more expensive than others. Some steps can be said to be only administrative in nature. Thus, their costs will probably be considerably less than investments in hardware, software, and personnel.

The examples in this section were developed by simply examining the heuristics in Chapter 8 and extending these heuristics to specific application domains. This analysis is qualitative, and no specific monetary costs were evaluated for each characteristic. This list is not meant to be comprehensive but rather a framework for investigating specific resilience features for any application domain.

11.4.1 Civil Infrastructure Application Domain

For almost any kind of disruption from terrorist attacks to hurricanes, the least costly characteristic would be the collaboration among various elements of the system. The heuristics that would apply include the knowledge between nodes heuristic, the *intent awareness* heuristic, and the *interelement impediment* heuristic, as discussed in Chapter 8. Assuming that the communications media are already in place—such as radios, landline telephones, mobile telephones—the process to achieve the heuristics are, for the most part, administrative. We discussed in Chapter 4, Case Histories, that these factors were in place in the New York Power Restoration case and largely absent in the Hurricane Katrina case.

To implement the *interelement impediment* heuristic, it would be necessary, first, to assure that administrative agreements are in place among all the nodes of the civil infrastructure system. These would include, at a minimum, the government, law enforcement, fire, power, medical, and transportation nodes. The second impediment to overcome would be any restriction on the free flow of information. These methods are well-known in the telecommunications domain. All the collaboration features would be minimal cost.

One of the most useful heuristics to examine from a cost point of view is the *functional redundancy* heuristic. The costs for keeping an adequate number of police, firefighters, and medical personnel on duty to handle a worst-case terrorist attack, earthquake, hurricane, or pandemic would be large. The alternative of first choice is probably to use resources from remote cities and states. However, according to Heisel and Weikel (2007), the planning for remote resources in the Southern California wildfires of 2007 was inadequate and resulted in shortages. Planning for the use of such remote resources may involve many details, such as the commonality of firefighting equipment and training in the fighting of rural fires, as opposed to urban fires. Although equipment and training may incur some costs, they are generally small compared with staffing each node to meet the worst-case situation.

Finally, there is the *reorganization* heuristic. This heuristic has to do with the ability of the civil infrastructure system to restructure itself in the event of a crisis. We observed a good example of this heuristic in the New York Power Restoration case. In this case, the New York Power system created an instant organization to manage the system of generators deployed after the twin towers attack. One can say that such reorganization is primarily administrative, and therefore the cost is low.

However, the most important finding is that resilience is much cheaper to implement for a civil infrastructure domain than protection. Perelman (2007, p. 25), for example, calls protective systems "hard" systems and resilient systems "soft" systems. For electrical power systems, for example, it is much less expensive to add redundant power lines than to protect each element of the system. Perelman (2007, p. 25) says, 'hard' technical systems were so capital-intensive, environmentally hazardous, and vulnerable to catastrophic break-down that they threatened to subtract more from the economy than they added. But alternative 'soft' technical systems were readily available that cost less and could be implemented more swiftly and safely."

11.4.2 Flood Protection Domain

Defra (2007) provides an insight into low-cost methods to improve the resilience of personal property—homes and businesses—in the face of floods in the United Kingdom. The study recommends an effort to inform the public about flood effects and preparedness. This topic is compatible with the *drift correction* heuristic, which requires advance notice of a disruption and action to deal with it.

The Defra (2007) report also suggests temporary inexpensive measures, such as floodguards, that would be put in place well before the flood. This measure would comply with the *margin* heuristic and would also comprise the correction aspect of the *drift correction* heuristic.

This report also concluded that employing the above measures without considering repair was not desirable. This conclusion is compatible with the *reparability* heuristic.

In a companion document, Defra (2007) recommends discouraging development in flood-prone areas. This recommendation is in agreement with the *functional redundancy* heuristic. This measure can be implemented by legislative action.

11.4.3 Aerospace Domain

With respect to the product itself—that is, the aircraft or spacecraft—the most well-known heuristic is the *physical redundancy* heuristic. Although this heuristic would be extremely expensive for major product subsystems, such as the engines, it is commonly invoked for lightweight components, such as electronics.

One cannot say that the Billings heuristics listed in Chapter 8, Resilience Architecting, are expensive because they are a one-time consideration. These have to do with the interaction between the operator and the automated system. The *human-in-control* heuristic is one example.

Among all domains, the *organizational planning* and *organizational decision-making* heuristics are critical. Although these heuristics may add some cost to the product-centered infrastructure cost, their benefit to the resilience of the product itself will be evident.

Finally, there are the risk and culture heuristics, such as the *small problem focus* heuristic and the *cultural change* heuristic. These heuristics are, for the most part, managerial and therefore do not add to the cost of the product. The Challenger and Columbia cases discussed in Chapter 4, Case Histories, support this conclusion.

11.5 RISK-BASED COSTING

Paté-Cornell (1990) and Leveson (2002) both advocate using risk as the primary measure of system resilience. Although their methods differ, the idea is basically the same, namely, that when risk is low, you know when you have done enough. Paté-Cornell uses probabilistic risk analysis (PRA), whereas Leveson uses a nonlinear, dynamic model developed at the Massachusetts

Institute of Technology (MIT). These models are discussed in Chapter 10, Measuring Resilience.

Regardless of the model, risk analysis involves quantifying both the likelihood (L) and the consequence (C). Because the values of both are judgmental, especially likelihood, there will most certainly be uncertainties in the values. Leveson points out that the values will depend largely on the job title of the estimator. The values assigned by managers, engineers, pilots, maintainers, and production personnel will most certainly differ. Paté-Cornell offers studies to show that this is true in real life.

One way to reduce this uncertainty is by independent reviews as discussed in Chapter 5, Culture, which offers several ways to overcome the cultural effects.

11.5.1 Will System Resilience Make Systems Cost More or Less?

The conventional wisdom is that system resilience will make systems cost more. That is, when all the aspects of the capabilities listed in Chapter 6, Capabilities, are implemented, costs will increase. For example, Chapter 6 calls for an increased level in independent reviews. Independent reviews are not free. Someone has to pay for them. All the other capabilities can be viewed as more detailed and hence more expensive.

M. Hamill (address to the Los Angeles Chapter of INCOSE on October 10, 2006), however, argues that system resilience will make systems cheaper. His argument is based on the premise that cost overruns result from the inefficiencies that arise from cultural problems, such as those listed in Chapter 5, Culture. This hypothesis is difficult to prove or disprove. Some initial estimates may be captured by assigning values to the cultural factors in the MIT risk model described by Leveson (2002). Even then, the potential savings would be only approximate.

In the end, the answer to this question can only be found by trying it. That is, if a major program were to have an intensive system resilience initiative and the program cost came down, then Hamill's theory would be validated. If that were to happen, then the case for initiating system resilience on all programs would be strong.

11.5.2 Cost Optimization

Chapter 10, Measuring Resilience mentions cost optimization. The question is: if you wanted to achieve the lowest level of risk for a given amount of money, how would you allocate the money? The answer lies in the ability to quantify risk/cost slopes, such as d(Risk)/d(Cost) individually for categories, such as test, requirements, safety, and so on. If data exist to estimate these slopes, then analyses called "portfolio optimizations" can tell you how much money to invest in each category.

11.5.3 Cost Control and Resilience

Probably the most difficult question an executive has to make is how much can cost be controlled without sacrificing resilience? Leary (2008) relates the struggles at NASA to decide how much cost to cut and what to do when cost overruns are experienced. Leary says that preflight tests were "scaled back" to save money. This step raises some tough questions: First and foremost, was resilience jeopardized by eliminating tests? Second, why were tests scheduled that were not needed? Finally, who made the decision to add or eliminate tests, administrators or technical experts? It is impossible to answer these questions, but some general guidelines can be related.

First, all costing should be bottom up and not top down. That means that the total cost of a program should be estimated by accumulating the costs to do each task in the work breakdown structure (WBS) rather than starting from a top-level total cost constraint.

Second, this is not to say that cost estimates should be accepted from the technical organizations without scrutiny. Each task should be examined carefully for its effect on resilience. Most importantly, each task should be examined for its contribution to resilience. Although this may seem obvious, the track record for doing this is not good.

Finally, if the total is more than the money available, then the viability of the program should be reexamined rather than scaling back and reducing resilience. Another option is to reduce the size and objectives of the program. For example, the duration of a mission can be cut.

11.6 CONCLUSION

One might ask: Why even discuss cost when cost cannot be quantified from a resilience point of view? Several answers to this question are available.

The first answer is that quantification is a serious cultural issue that needs to be addressed and overcome. It is particularly endemic to engineering organizations in which everything is thought to be quantifiable. This issue is addressed in the *design focus* paradigm of Chapter 5, Culture.

Second, even if the quantification issue is resolved, there is the real-life fact of money. Everything costs money. The customer, whether it is a commercial or government customer, wants to know how much the system will cost. The simple answer to this fact is that builders of systems will have to become better at figuring resilience into the cost of a system. In the construction of large systems, overruns have become a fact of life, and customers, to some degree, have come to expect them. Moody et al. (1997) showed from NASA data that the more systems engineering and other upfront planning is put into a program, the *less* the overrun will be. As resilience is better understood, the better cost estimating should get. Finally, there is the theory advanced by Hamill (2006), as mentioned above, that when you address resilience, the cost will actually go down. This theory has yet to be validated, but it does present a hypothesis for future study.

So, to answer the initial question, how do you know when you have invested enough, you have invested enough when the risk is low. You also have to answer the question: how do you know when the risk is low? Once again, we are in the initial phases of more advanced risk analysis. We have discussed the approaches of Leveson (2002) and Paté-Cornell (1990) in Chapter 10, Measuring Resilience. There is also the perspective of Epstein (2006) discussed in Chapter 6, Capabilities. And last, but not least, it is necessary to conquer the risk denial paradigm discussed in Chapter 5, Culture.

So when these methodologies become more mature, we might have a handle on the question: How much will it cost?

11.7 FURTHER EXPLORATION

In your research of case studies, make a list of those cases in which cost as a constraining factor contributed to the system failure. Provide details.

Chapter 12

Implementation

The preceding chapters have laid the groundwork, the rationale, and the scientific basis for the creation of a resilient system that will maximize resilience. This chapter will summarize the attributes of that infrastructure and the recommended steps that should be taken to realize that system. Obviously, these steps can only be accomplished by high-level managers, such as program managers, or by a group empowered to make such decisions. Regardless of the person or group making these decisions, they should have the authority that transcends technical, managerial, and contractual boundaries. This authority is especially important when considering human systems (government organizations, for example) or product systems (hardware and software).

The following system descriptions are not prescriptive; that is, it is not intended to give a precise definition of what the system is but rather to describe what characteristics would result in a resilient system, both functionally and organizationally.

12.1 IMPLEMENTING RESILIENCE CAPABILITIES

Capabilities are the bricks in the house of resilience. They are the processes and procedures that almost all organizations employ. Chapter 6, Capabilities, lists the capabilities that are most critical to system resilience. Although most of these capabilities are common in many organizations, it is necessary to determine whether all these capabilities are actually documented processes or procedures within the organization. Most organizations would have, for example, a

Architecting Resilient Systems: Accident Avoidance and Survival and Recovery from Disruptions.
By Scott Jackson

requirements process especially for product systems, to assure that nothing is actually built before the requirements have been established for them. Many organizations might not, for example, have a process for decision making or for business continuity planning. In any event, most resilient organizations will employ all these capabilities.

12.1.1 Architecting

Many organizations do not have a capability known as architecting. Chapter 8, Resilience Architecting, addresses this question. When considering architecting, it should be kept in mind that some systems are product systems (hardware and software) and others are human systems (government agencies, for example). In each case, there needs to be an organizational element with the expertise to create the resilience attributes of capacity, flexibility, tolerance, and interelement collaboration.

Each organization within the resilience infrastructure should create the position of systems architect. This position may have different names, but the result should be the same. The systems architect, using the principles of systems architecting, of course, will first address the subject of the organizational architecture. He or she will, using the heuristics of Rechtin (2000) and Chapter 8, Resilience Architecting, create an organizational architecture deemed to be in concert with the principles of resilience. This architecture will support the attributes of resilience as laid out in Chapter 8.

A key task of the systems architect is to assure that the organizational element he or she represents is an integral part of the infrastructure system of systems as described in Chapter 6, Capabilities, and Chapter 7, Infrastructure.

For product systems, the systems architect will employ the heuristics described by Rechtin (1991). In both cases, the heuristics should support the attributes of resilience described in Chapter 8.

12.1.2 Analytic Capabilities

For a system to be resilient, it should first be dependable. For product-centered enterprises, that is, enterprises whose main function it is to develop, produce, and deploy technological systems, such as aircraft and spacecraft, the top-level group would oversee the principal product-centered functions, especially those discussed in Chapter 6, Capabilities. These include requirements management, verification, risk management, interface management, and configuration management. It would also include other functions, such as system safety and reliability. Although these functions come under the general category of *durability*, they are essential for the survival aspect of resilience, that is, Phase 2 of resilience as described in Chapter 1, On Resilience. In addition, they would address the more advanced aspects of subjects such as safety and risk, as described in Chapter 6.

12.1.3 Considerations Regarding the Implementation of Advanced Resilience Capabilities

The question arises as to when the advanced resilience capabilities of Chapter 6, Capabilities, should be employed. Should they be used for all types of systems under all conditions? Should they be used for disruptions whose probabilities are exceedingly small? The answer is probably "no" to both these questions. In summary, the probability of a disruption is not the only basis for implementing resilience capabilities.

Following are some guidelines as to when these methods should be used.

12.1.3.1 Low or Negligible Cost Situations. We saw in Chapter 11, Cost, that some implementation features can be implemented at little or no cost. Sometimes, these opportunities are available by chance, that is, the circumstances of a particular scenario. For example, in the Columbia accident discussed in Chapter 4, Case Histories, visual surveillance was available at no cost or negligible cost. Unfortunately, this offer was declined.

Often, communications capability is available without the need to buy new equipment. This capability is an essential ingredient in the interelement collaboration discussed in Chapter 8, Resilience Architecting. Interelement collaboration can also be enhanced by simple reorganization of the infrastructure as discussed in Chapter 17, Infrastructure. Reorganization may be achieved at little or no cost.

For large technological systems, such as spacecraft and commercial aircraft, high-fidelity simulations have become commonplace. Tools, such as finite-element analysis (FEA) and computational fluid dynamics (CFD), have been available for many years. Hence, it is conceivable that these types of tools could be extended, which would have anticipated the hidden interaction experienced on the Mars Polar Lander, as described in Chapter 3. The development of such tools is normally amortized over many such programs so that the cost to an individual program would be small. To address the effects of undesirable interactions discussed in Chapter 3, Disruptions, these tools would need to address such interactions as heat, vibration, electromagnetic interference (EMI), the functional effects of software outputs, and other such parameters.

12.1.3.2 High-Consequence and High-Risk Systems. Weick and Sutcliffe (2001) make the case that systems, such as nuclear power plants and aircraft carriers, deserve more detailed attention than other systems. The same could be said for other systems, such as space vehicles. Focusing on small problems is the central emphasis of systems such as these. The reason is simple: A catastrophic accident resulting from the smallest disruption can lead to the loss of many lives and substantial destruction of property. The Concorde accident discussed in Chapter 4 is one example of a catastrophic accident resulting from a small disruption, a small piece of metal on the runway.

12.1.3.3 High Benefit-to-Cost Ratio. Chapter 11, Cost, also shows that in certain industries, such as commercial aircraft, quantitative methods have been used to determine whether system improvements are indeed warranted. To the extent that such analyses are valid, this methodology does provide a gauge for determining the depth of analysis required to make such decisions.

12.2 IMPLEMENTING AN INFRASTRUCTURE

Creating an infrastructure may be the most difficult of all aspects of achieving system resilience. Following are some features of a robust organizational infrastructure for system resilience.

12.2.1 The Enterprise Infrastructure

In this book, the term *enterprise* is used to describe the set of all organizations that may contribute to system resilience. Getting all the elements of the enterprise to act as a whole requires actions on the part of all the elements. The reasons these elements need to act as a whole is that failures can result from causes that are areas of mutual responsibility. Chapter 4, Case Histories, cites many such cases.

So, what is the mechanism for such collaboration? There are many possibilities. But at a minimum, an agreement or a memorandum of understanding (MOU) would be one method of achieving this goal. The basic assumption behind such an agreement is that there are areas of endeavor that go beyond contractual expectations. It is expected, for example, that a developer will have an infrastructure in place that will enable safe systems.

In addition to the agreement, there are possibilities for an organizational element to help achieve the goal. For example, a coordinating group consisting of customer, developer, and supplier representatives may be established. The degree of authority vested in this group is up to the organizations involved. However, just an agreement to address areas of mutual responsibility will be a major step.

There are differences among a civil infrastructure enterprise, a private enterprise, a product-centered enterprise, and a public enterprise. The heuristics of resilience apply to all types of enterprises.

12.2.1.1 Civil Infrastructure Enterprise. The civil infrastructure enterprise is not a unified whole. It can either be a system of systems, as defined in Chapter 2, Systems Resilience and Related Concepts, or it can be a federation of systems. In the system of system, on the one hand all the elements of the system are under some control of a common entity, such as a regional authority. In the case of the federated systems, each element of the system of systems is known as a stand-alone system. An example of a stand-alone system of a Fire Protection Infrastructure System is discussed in Appendix A to this book. The Fire Protection Infrastructure System is also discussed in the context of a system of systems.

In any case, a stand-alone system should institute the capabilities of a resilient system without the aid of a larger entity. Appendix A suggests that the Fire Protection Infrastructure System should institute the agreements with outside entities, such as government agencies, to establish the interelement collaboration essential to resilience.

In the case of systems of systems, the authoritative element of the system, be it a regional or national entity, is responsible for establishing these capabilities. In many cases, however, it has been found that there was no entity that had the responsibility, or was willing to take the responsibility, for assuring the resilience of the entire system. Two examples are the Metrolink 111 accident near Los Angeles in 2008 and the Hurricane Katrina infrastructure in New Orleans in 2005. Both of these incidents are discussed in Chapter 4, Case Histories. The Metrolink 111 accident is also discussed in Appendix B, A Resilience Analysis of Metrolink 111.

12.2.1.2 Private Enterprise. The term *private enterprise* is used here to designate any organization created to develop a product or service. Government organizations are excluded. Private enterprises may be government contractors for either defense or space purposes. Although a private enterprise may have more autonomy than a civil infrastructure element, it has its own challenges with respect to interelement collaboration, both internal and external, and to creating a resilience-friendly organization. The paragraphs below will address the implementation of system resilience in the private enterprise domain.

12.2.1.3 Product-Centered Enterprises. These enterprises are normally a type of private enterprises. They can either be systems developed for the government, such as spacecraft, or for the public, such as commercial aircraft and automobiles. The capabilities for developing these technological systems are generally well defined, such as requirements analysis, verification, interface management, and so forth. These capabilities are sometimes known as analytic methods, as defined in Chapter 6, Capabilities.

In some fields, such as commercial aircraft, the use of heuristics is already well established. The Billings heuristics discussed in Chapter 8, Resilience Architecting, are a standard set of guidelines for design. It can be concluded, then, that the use of such heuristics in other fields is an advisable path. This use would particularly apply to fields in which there are human–machine interfaces. With the rapid increase in information technology, this conclusion could apply to almost any domain, even civil infrastructure domains.

The capability that needs to be emphasized in product-centered enterprises is governance, as discussed in Chapter 9, Governance. We saw in Chapter 4, Case Histories, that failures occurred at a level of detail that is common in practice.

The English-metric confusion that resulted from a faulty interface process on the Mars Climate Orbiter is an example.

The resilience community is also in agreement that the role of safety in product-centered enterprises is one that should be expanded. Woods (2006), for example, recommends a stronger role for safety in organizational issues. That is, safety should be involved in issues beyond the design of the product system.

12.2.1.4 Public Enterprise. The term *public enterprise* refers primarily to government agencies that have a key role in system resilience, especially with respect to overseeing a multitude of lower level agencies and private agencies and enterprises.

Among these enterprises there are, for example, regional authorities, such as the New York Port Authority. They can include security agencies, such as the Federal Bureau of Investigation (FBI), the Federal Emergency Management Agency (FEMA), or the National Guard. In some cases, the enterprises have enforcement authority and resources. In other cases, they have only coordinating power.

Regardless of the authority they may have, public enterprises have a unique role in civil infrastructure resilience. Primarily, they have the roles of coordinators and planners. Their role is to determine what resources are required where and when. As pointed out in Chapter 10, Cost, the least costly resilience capability is communication, especially if radios and telephones are already in place.

The first step for these enterprises is a plan. This plan will include a comprehensive list of resources of all civil infrastructure elements within their jurisdiction. These elements will include hospitals, police departments, fire departments, power companies, transportation entities, and so on. Within each of these elements will be listed specific assets, such as personnel, equipment, vehicles, and so on. This plan will also include points of contact and multiple ways of contacting key personnel, in accordance with the functional redundancy heuristic of Chapter 8, Resilience Architecting.

The final element of the plan is a series of agreements designed to make the system resilient to any disruption. These agreements will spell out who and what resources are to be made available to other elements, how communications will be conducted, and so on. In the end, the objective of the plan is to make the public enterprise system of systems function as a single system in the face of a disruption.

12.2.1.5 The Internal Infrastructure. Any organization, whether it is a civil infrastructure system, a private enterprise system, or a public enterprise, needs to have a person or group responsible for the resilience of that organization. Whether this group has authority or is just advisory will vary with the organization. If the organization is large, it may have representatives from various groups within the organization. For smaller organizations, such as a police department, a single person may fulfill this role.

In some cases, the resilience group may have a degree of independence from the entire organization and may have the authority to make decisions independent of the organization. In an aerospace organization, for example, the group may have the authority to deny launch of a space vehicle.

The resilience group should be broad based; that is, it cannot oversee just the technical organizations but rather all managerial and technical organizations. It should report to the highest level person in the organization. This position will afford it the maximum level of communication with the leader.

Another role of the system resilience group is its external liaison role. This organization will be the primary interface with both the customer and the suppliers. It will be responsible for any agreements with these external organizations. If the organization is an element under the surveillance of a *public enterprise*, as described above, this liaison will be responsible for this coordination and collaboration with the public enterprise.

For example, an organization may have a communications process. But the communications process may be high level and apply only to external communications. Hence, the communications process may need to be made more robust to address the detailed communications paths that are necessary for system resilience.

12.3 IMPLEMENTING A RISK PROCESS

It is not that risk is not studied and written about extensively. It is not that organizations do not have risk processes. But are risk and its subtleties understood? Is risk taken seriously? Is risk management considered a *pro forma* process? These questions are to be answered in the implementation of risk.

Many organizations employ the standard risk model discussed in Chapter 6, Capabilities. This model assumes that events with a serious consequence and a high likelihood of occurrence are the ones to receive attention. This model is a good start. But the implementation of this model raises questions. For example, are the risks outside the control of the organization identified? The implementation of this model will be compromised if basic questions like these are not answered.

Going beyond the basic model, there is the dynamic-nonlinear model of Leveson et al. (2006). Although only large organizations, such as the National Aeronautic and Space Administration NASA, may be able to afford to implement this model, it may add value in analyzing risks that would have otherwise been neglected.

Also, there is the observation of Epstein (2006) who states that most accidents occur in regions of low probability. Although these events may be difficult to analyze, Epstein's observation raises the bar on risk management. That is to say, in the implementation of risk an organization should look beyond the high-likelihood events.

For small organizations, such as fire and police departments, who may not have the resources to have dedicated risk tools or departments, recognition of risks in these domains remains a necessity, even on an informal basis. These organizations have national and international societies who can be helpful in the identification and treatment of risks.

The linchpin of risk implementation is culture. We saw in the Challenger, Columbia, and Texas City–2005 accidents discussed in Chapter 4, Case Histories, that culture and the resulting neglect of risk were a major factor. The discussion below addresses the implementation of cultural initiatives.

12.4 IMPLEMENTING A CONTRACTUAL SYSTEM

The implications of system resilience on the contracting process are not insignificant. It is not that the mechanisms do not exist today, but there will need to be a change of emphasis. Let us discuss three types of systems with respect to contracts.

12.4.1 Infrastructure Systems

Because infrastructure systems do not involve a product, they would require a different approach. Mechanisms would be, for example, internal guidebooks, directives, and memoranda of understanding (MOUs).

Police departments, fire departments, or government agencies could have a guidebook to help them develop their own infrastructure and procedures to maximize resilience. This guidebook would, of course, address their relations with other organizations, that is, infrastructure nodes. Traditional contractual mechanisms will apply for such products as trucks and other equipment. However, the contractual relationship between higher level nodes and lower level nodes will evolve into a more defined set of mechanisms.

Infrastructure nodes with subnodes could develop directives to assure that their subnodes complied with the principles of system resilience. Examples are organizations at the national, state, provincial, county, or city levels. As a generalization, it can be concluded that higher level nodes, such as national or state levels, have more responsibility for planning and coordination among lower level nodes.

Finally, any infrastructure node that is a system within a system of systems can develop MOUs with sibling nodes. These MOUs would establish the necessary relationships and actions that should take place in the event of a disruption, such as a hurricane or terrorist attack.

Many metropolitan areas, such as Los Angeles and New York, have regional authorities to address such issues. These authorities are the logical initiators of such guidebooks, directives, and MOUs.

12.4.2 Technological Systems

For technological systems, even systems with a considerable number of human components, current contracting practices are heavily reliant on specifications. In the resilience environment, this method should give way to a less quantitative approach to contracting. Suppliers, for example, would need to present different designs, which would be evaluated for their resilience. This approach could be captured in a statement of work (SOW) rather than a specification.

12.5 IMPLEMENTING A MEASUREMENT SYSTEM

Another aspect of system resilience capabilities is that they should be subject to change based on metrics and other input data. Chapter 10, Measuring Resilience, for example, provides predictive data on the effectiveness of the infrastructure and of the individual capabilities themselves. The measurement process determines whether the infrastructure system is robust or not. Based on these data, process changes or even organizational changes can be effected.

The next feature of a resilient system is a system resilience measurement function, as described in Chapter 10, Measuring Resilience. It does not matter where in the organization this function resides as long as it has the capability to measure key parameters, analyze the results, and provide guidance for the improvement of the system capabilities, as described in Chapter 6, Capabilities.

The key function of the measurements is to provide indicators of possible lack of resilience. For this reason, especially in the conceptual design and development phases, the measurements should be predictive. As the product is delivered and enters into the operational phase, the measurements may be reactive, that is, they can measure defects in the delivered product and actual failures. Statistical analyses will be used to show whether these measurements, either predictive or reactive, are indicators of future failures.

Another feature of the measurements is that they should focus on quality rather than on schedule or cost, the latter being considered constraints rather than indicators of system resilience.

The system resilience team will use these measurements for making its recommendations for element improvements. The implementation of a measurement system can be divided into two broad categories: development metrics and operational metrics. The two following sections will not suggest specific metrics to measure these categories; they will discuss some broad categories that have been used in the past and the success or difficulty of these approaches.

12.5.1 The Implementation of Development Metrics

Development metrics ask the question: Has the system been designed, tested, and produced in the best possible way? This does not matter whether the system is a hospital, a fire department, or a space vehicle. This type of metric is the more difficult of the two categories of metrics. First, it makes the assumption that there is a right way and a wrong way to develop a system. Second, it assumes that if a process is followed, a resilient system will result. Neither of these assumptions is categorically correct; it can only be said that experience has shown that a better system will result. There is no guarantee.

So how does a metrics system measure heuristics? The only answer is to consider the employment of heuristics to be a Boolean metric, that is, a yes/no metric. Either they were considered or they were not. There is no middle ground. So to satisfy the *functional redundancy* heuristic for a Fire Protection Infrastructure System, for example, as illustrated in Appendix A, the question would have to ask: Are there multiple ways to suppress a fire? The answer is either yes or no. If it is yes, then the system could be considered more resilient than one with only one way to suppress a fire.

The CMMI® (2006) metric system does treat process metrics in a non-Boolean way by rating each process on a 1-to-5 scale with each level representing a different level of process quality. A level of 1, for example, might represent an *ad hoc* nondocumented level of compliance, where a level of 5 might represent a rigorous and universal level of compliance with continuous improvement. This method is widely used in aerospace and other large organizations.

There are, however, some objective measures in the development of a system. For example, how many flaws in software code are routinely discovered? How many tests fail? These measures can be applied to human-intensive systems, such as fire-fighting systems and hospitals. The level of training of personnel, for example, can be measured.

As discussed in Chapter 6, Capabilities, Leveson et al. (2006) use risk as a metric. Although this metric is logical, its execution is somewhat subjective. The use of risk as a metric involves the estimation of probabilities of future events. Some of the probabilities shown by Leveson et al. involve probabilities associated with human behavior. So, although the use of risk as a metric is logical, it should be done with caution. In addition, its use is dependent on solving the cultural issues associated with risk, as discussed in the next paragraph.

The most difficult development metric of all is the measurement of culture. We discussed in Chapter 4, Case Histories, that culture was a major factor in the Challenger, Columbia and Texas City–2005 accidents. So, how can culture be measured? This question is for organizational psychologists, but one of the more popular methods is the use of personnel questionnaires. Questions, such as "Is the autonomous reporting system really autonomous?" can be very revealing. How culture can be improved is a more difficult question. Chapter 5, Culture, provides some alternatives.

Mallak (1998) provides one example of a study to measure culture in the health care industry. In this study, two cultural factors, goal-directed solution seeking and critical understandings, among others, are quantified and measured. Among all these factors, goal-directed solution seeking was found to be the most important.

12.5.2 The Implementation of Operational Metrics

Operational metrics are much easier to collect, and their results are easier to interpret, but the major disadvantage is that by this time it may be too late. If major catastrophes have already occurred, then the best hope is to fix the system to reduce the risk in the future or to use the lessons learned for future systems.

One of the most promising approaches is the use of defects and near misses as predictors of brittleness, which is the opposite of resilience. We saw in Chapter 10, Measuring Resilience, that Wright and Van der Schaaf (2004) found a strong correlation between the causes of defects and fatalities in the Scottish Railways study.

One advantage of the correlation approach is that it includes all defects and fatalities, not just the ones that were repeats of past accidents. Hence, it is one way of taking into account unpredicted disruptions because some events in the study may have been unpredicted.

Hospitals are one of the major sources of operational metric data. Okada (2003), for example, analyzes parameters for human error potential (HEP) that contribute to errors in hospitals.

Once again, the value of these metrics is that they can be fed back into the system to increase resilience in the current system or to architect future systems.

12.6 IMPLEMENTING GOVERNANCE

The system resilience organization has three functions with regard to overseeing the system resilience efforts on a program.

First, the system resilience organization is responsible for reviewing all current processes, the metrics associated with those processes, and existing corrective action efforts on a program. From this review will come recommendations for improving processes, adding new processes, and even modifying the organization or contractual aspects of the program, either with the customer or with suppliers. One process to be initiated by the system-resilience organization is the cultural initiative described below and in Chapter 5, Culture.

Second, the system-resilience program team will function as a team or committee to perform these reviews. Overseeing this team will be a basic function of the system-resilience organization.

Finally, the system-resilience program team will oversee independent reviews, one of the basic capabilities to ensure system resilience. As discussed in Chapter 6, Governance, reviews can either be scheduled on conducted or an as needed basis.

12.7 IMPLEMENTING A RESILIENT CULTURE

Chapter 5, Culture, shows that numerous cultural beliefs, or paradigms, can be detrimental to system resilience. An example of such beliefs is the failure of many organizations to have a robust risk-management process, not because there is no process or risk-management tool, but because cultural barriers prevent risk management from being treated with the importance it deserves.

Chapter 5 also shows that the traditional methods of training and executive direction have had limited effectiveness in dealing with these paradigms. Alternative methods, for example, communities of practice, are suggested as potential alternatives for changing paradigms.

In any event, Chapter 5 suggests that a cultural change initiative needs to be part of any enterprise. It is not particularly important what organization within the infrastructure is responsible for implementing the cultural change initiative. It may be within the traditional training organization, even though that name may not be appropriate for this initiative. Cultural change is a key element within the system-resilience infrastructure.

A key question with regard to culture change is where the culture initiative starts. The "new model" described in Figure 5.3 shows that any culture change starts with the executive. If the executive believes in the principle of self-discovery, then this method has some validity. If the executive is not an authority on culture, then the results may be less than satisfactory. However, there is no reason the initiative cannot begin with a group of dedicated individuals within the organization. This method will work if the executive appreciates the work of the individuals. Either way, cultural change is possible.

Chapter 5 also shows that an alternative way to deal with the culture issue is to manage culture rather than to change culture. In the near term, managing culture may be the most practical approach. As a matter of fact, this approach is taken in most cases for which culture has been found to be a major issue. The Baker commission (2007), for example, recommended this approach following the Texas City–2005 accident as discussed in Chapter 4, Case Histories.

The system-resilience infrastructure does not imply just one organizational element of the infrastructure, for example, the developer organization. Rather, all organizational elements should cooperate on this initiative, for example, developers, customers, suppliers, maintainers, and regulatory agencies. The system-resilience node, discussed above, will have primary responsibility assuring that the cultural initiative exists in all infrastructure organizations.

12.8 IMPLEMENTING A COST PLAN

Chapter 11, Cost, deals with the most difficult system resilience question: how much will system resilience cost? Despite this difficulty, cost decisions will be made, and therefore, cost analysis will be part of the system-resilience functional infrastructure. The reasons for this difficulty are twofold: First, the likelihood of system resilience is not quantifiable and, second, there are no known goals to measure system resilience. Nevertheless, the following characteristics of a system resilience cost decision process can be defined.

The system-resilience organization, as described above, will have the primary responsibility for making recommendations to program management pertaining to cost. This organization will use predictive metrics as the primary means of determining whether additional costs need to be expended as described in Chapter 10, Measuring Resilience. The primary metric in cost decisions is risk. Leveson's (1995) risk analysis provides a clue as to how this question can be answered. Program management will have the final responsibility for cost decisions.

12.9 SUMMARY

The implementation approaches described above may depart from current practices in many domains. However, it is expected that these practices will become more universal as time goes on.

12.10 FURTHER EXPLORATION

Compare how system resilience would be implemented in a human-intensive system of systems, such as a hospital, with how it would be implemented in a product focused infrastructure. Provide a detailed discussion and diagrams.

Chapter **13**

A Summary of Themes

This book has discussed many aspects of system resilience from which one may infer certain themes. The following section is a summary of those themes.

Theme Number 1: System resilience is dependent on factors beyond the technical, which include managerial, infrastructure, and cultural factors.

Although this theme may be apparent to anyone studying system resilience, it is a hard concept to grasp for many scientists and engineers whose experience is focused on quantitative technical analysis. Nevertheless, extensive research, as documented throughout this book, bears out this fundamental truth. We have discussed many examples in which factors, such as poor communications and decision making, contributed to a major accident. The North Sea Oil Platform disaster, as discussed by Paté-Cornell (1990), is but one example. Hurricane Katrina as discussed in Chapter 4, Case Histories, is an example of a brittle infrastructure. The Metrolink 111 accident also discussed in Chapter 4 and in Appendix B is another example. Challenger, as discussed by Vaughn (1996), and Columbia, as discussed by the Columbia Accident Investigation Board (2003), are examples of accidents for which culture was a major factor.

Theme Number 2: At a minimum, both technical and managerial capabilities should be executed as a precondition for system resilience.

It is easy to conclude that because such a wide variety of sources is the root causes of accidents, then technical and managerial processes take a back seat. Nothing could be further from the truth. On the contrary, organizations should execute technical and managerial processes as if they were a single process. Second, they should be conducted with a rigor not common in industry today.

A subtheme is that there should be no dividing line between management and technical capabilities. In the words of Feynman (1988), there should be a "common interest" between the two areas.

Theme Number 3: The system-resilience infrastructure is a system of systems that should function as an integrated whole.

This theme may also consist of resistance in government and industries that focus on contract-based obligations and seldom look beyond their own organizational boundaries. This is not to say that contracts should be ignored but that contracts should be written to foster cooperation rather than line drawing.

The ValuJet accident, as described, for example, by Chiles (2002, pp. 150–155, 310) is just one example of how multiple organizational boundaries only serve to erode resilience. In this case, an airline outsourced a maintenance organization, which was three degrees of separation from the aircraft developer, to perform activities counter to the specified safety practices. The result was a disaster.

Another feature of the infrastructure that is sometimes hard to understand is the fact that the infrastructure is a system at all. For most traditional systems engineers, the system is the product system, that is to say, the aircraft or spacecraft. However, from the resilience point of view, the analyst needs to understand that the infrastructure is a system itself that needs to operate as a whole. When it is considered that this infrastructure contains the customer, the developer, the suppliers, and many other elements, the importance of this concept becomes more apparent.

In some cases, the infrastructure is the only system of interest; that is to say, there is no product system. For example, the infrastructure that had the job of protecting New Orleans from a hurricane is, for the most part, a human system. The elements consist of agencies, such as the police, fire department, hospitals, and so on. The concept of the infrastructure as a system is no less valid in this instance.

Once the analyst has internalized the infrastructure as a system concept, he or she should then tackle the concept of a system of systems. Within this concept, the developer, the customer, and the suppliers are systems themselves that should act together to achieve resilience of the entire system. In the case of the hurricane infrastructure, the police department, fire department, and so on are systems in themselves that need to work together. The key feature of a system of systems is that it has little central control. All the elements should work together in a collaboration to achieve resilience. This system of systems aspects leads to complexity and the likelihood that disruptions will occur. The heuristics of Chapter 8, Resilience Architecting, increase the likelihood that the system will avoid, survive, and recover from these disruptions.

The two most notable cases of brittle systems of systems discussed in this book are Hurricane Katrina and the Metrolink 111 accident, which are both discussed in Chapter 4, Case Histories. The Metrolink 111 accident is also discussed in Appendix B. The responsibility for achieving control of such systems of systems in these cases rests, for the most part, on governmental action, which may be difficult to achieve.

Theme Number 4: System resilience can only be realized within a resilience-focused cultural environment.

It is important not to underestimate the importance of culture. The Columbia Accident Investigation Board report (2003), for example, states that culture is at least as important as the technical factors. However, culture cannot be considered as a vague unmovable object. If culture is a problem, then organizations need to deal with it at the source or, at a minimum, eliminate the factors that spawn negative cultural paradigms.

Theme Number 5: Case histories provide clues of common root causes of accidents and the needed capabilities for future systems.

The most striking value of case histories is the finding that many catastrophes have common root causes. Examples are a lack of attention to risk, deficient maintenance, and a lack of collaboration among elements of the system. These findings lead to the logical conclusion that specific enhancements in certain capabilities are required to improve resilience.

Of almost equal importance is the finding that the root causes of some catastrophes are beyond the knowledge and predictability of the designers of the original systems. The treatment of these root causes calls for, in some cases, an increased scope of standard systems-engineering processes. In other cases, in which disruptions are completely unpredictable, adaptable systems are required, as discussed below in Theme Number 7.

Theme Number 6: Minor defects and near misses provide statistical measures of system resilience to major accidents and a methodology for system improvement.

Researchers have noted that in almost all disasters, the accidents were preceded by defects and near misses that might have been used as warning signs of a less than resilient system. This observation received some support by the findings of Wright and Van der Schaaf (2004), who showed a strong statistical correlation between the incidence of defects and the incidence of fatalities. These findings argue for increased research into these phenomena.

The logical question is as follows: What can be done with these data? The obvious answer is that the management of these systems can apply added capability where deficiencies exist.

A second answer is that management can determine the optimum allocation of funding to different aspects of the system and can apply funding in accordance with this optimization. This optimization is not an easy matter. It is not known whether such an optimization has ever been attempted before in a resilience context. If not, now is the time.

Theme Number 7: System adaptability is required for resilience to unpredicted disruptions.

The final question is, as discussed in Theme Number 5, how can a system be designed that will be resilient to disasters and to disruptions? The answer is that the system should be adaptable. Woods (2006b), for example, provides some guidance as to what the characteristics of an adaptable system might be. The unanswered questions include, for example, what are the design practices that will result in an adaptable system? The second question is as follows: Will an adaptable system be more expensive than another system?

Probably so, but not necessarily so. All these questions beg for even more answers.

Theme Number 8: Certain aspects of resilience are quantitative and others are not. The nonquantitative aspects can either be optimistic or pessimistic.

Chapter 10, Measuring Resilience, shows that certain aspects of resilience are quantitative and measurable. The quantitative aspects include reliability (especially for technological systems), system safety, and capacity. The qualitative aspects include unpredicted disruptions, heuristics, and—in many cases—human error. A system whose resilience is evaluated on the basis of the quantitative measures alone will be optimistic with respect to human error and unpredicted disruption, and it will be pessimistic with respect to heuristics.

Theme Number 9: The cost of resilience is linked directly to the ability to measure resilience. Hence, cost estimates will be either pessimistic or optimistic in accordance with the various aspects of resilience.

Chapter 11, Cost, showed by logic that the cost of resilience can only be gauged against quantitative measures of resilience. Hence, any cost estimate arrived at by the quantitative measures identified in Chapter 10, Measuring Resilience, will be optimistic with respect to human error and unpredicted disruptions, and it will be pessimistic with respect to heuristics.

Theme Number 10: The conclusions in this book are based on scientific observations from many fields.

Even though these conclusions and the practices that result from them may not be practiced in many domains at this time, they are based on sound observations. It is expected, that—with time—these practices will become accepted in the way that the Six Sigma and Lean processes have been accepted.

The field in which most of the principles of this book are currently practiced is cognitive engineering. The heuristics of Billings (1997), for example, which apply to the interaction between humans and machines, and have been applied mostly to commercial aircraft design, are applicable to many domains.

Theme Number 11: Resilience is dependent on at least three considerations: adaptability, risk, and culture.

In Chapter 8, Resilience Architecting, we discussed that the *design* of a system either human or technological should have four adaptability attributes: capacity, flexibility, tolerance, and interelement collaboration. Beyond adaptability, the infrastructure system, either stand alone or product centered, should be strongly risk based and culturally focused on resilience. Other considerations of importance are the environments listed in Chapter 2, System Resilience and Related Concepts. They include the economic environment, the regulatory environment, the contractual environment, the political environment, the organizational environment, and the geopolitical environment.

Theme Number 12: Resilience considerations are complementary.

This theme states that even if all considerations and attributes cannot be implemented, resilience is gained by each one individually. For example, a system that has a deficient capacity still gains in resilience by having a strong risk-based organization.

Theme Number 13: Before a system can be resilient, it should first be dependable.

It might be assumed that resilience only has to do with the "soft" or intuitive heuristics in Chapter 8, Resilience Architecting. The *human backup* heuristic is one example. On the contrary, it is not possible to recover from a disruption without first surviving it, and survival is the second phase of resilience as we discussed in Chapter 2, System Resilience and Related Concepts. To survive a disruption, a system should first absorb it as reflected in the *absorption* heuristic in Chapter 8. Absorption and survival are dependent, to varying degrees, on traditional analytic methods, such as reliability, safety, and design methods that assure, first, that the system has the capacity to survive the disruption.

It might also be assumed that terms such as reliability, safety, and design apply only to technological systems. On the contrary, human-intensive systems, such as civil infrastructure systems, should also be capable of absorbing disruptions.

Theme Number 14: Although humans may be the source of disruptions, the responsibility for the resilience of a system lies with the system owner.

Although it may be tempting to blame a human, such as an operator, for a disaster, it is rare that an individual either making an erroneous decision or performing an unwise act is responsible for the failure of an entire system. That responsibility lies with the system architect and the entity responsible for implementing the architecture of the system. As discussed in Chapter 4, Case Histories, for example, the train driver who may have been text messaging and missed the signal in the Metrolink 111 accident bears a responsibility for his individual act. However, the responsibility for creating functional redundancy in the system lies with a higher level authority. This authority can be called the system owner.

Theme Number 15: A disruption is generally not the root cause of an accident.

It is a common misperception that disruptions are the root cause of accidents. It is believed, for example, that a component failure or human error may be the root cause. It has been the theme throughout this book that the root cause of most accidents is brittleness, that is, the lack of resilience. When systems failed, it has been, for the most part, the inability of the system to anticipate the disruption, survive the disruption, or recover from the disruption.

In the case of Apollo 13, for example, the original disruption may have a reliability failure. However, it was the resilience of the entire system that allowed the crew to survive. In the cases of Helios 522, Mars Polar Lander, and Nagoya, it was the lack of resilience that eventually led to the catastrophic results.

Theme Number 16: A system may fail because of the interaction between components that operate as designed.

There is another common belief that a component, or another agent, such as a human operator, must fail to cause a disruption that will lead to a catastrophe. Leveson (2002), for example, states that this is a common assumption in traditional safety analysis. In several cases discussed in Chapter 4, Case Histories, the unanticipated interaction between the components resulted in the ultimate catastrophe. The Mars Polar Lander was an example. With diligence and detailed analysis, these interactions can sometimes, but not always, be anticipated and dealt with.

Theme Number 17: System brittleness often originates at the interfaces among system elements.

This theme applies to any type of system, either technological or systems with humans. For example, in the ValuJet case discussed in Chapter 4, Case Histories, the lack of communication and cooperation among loosely connected system elements was at the root of multiple accidents. This thesis is complementary with Theme Number 3 on infrastructure vulnerabilities and Theme 16 on the interaction among components.

Theme Number 18: The probability of a disruption is not the only consideration in the determination of what capabilities should be employed to make a system resilient.

Chapter 12, Implementation, provides guidance on what considerations should be taken into account when deciding on the depth of analysis and on the extent of the implementation needed to make a system resilient. At a minimum, these considerations include the cost of the implementation, the consequence and risk of a particular system, and the benefit–cost ratio of implementation.

13.1 FURTHER EXPLORATION

Take each theme and relate it to one or more case studies with details of how that theme would be applied in enhancing systems resilience.

Chapter **14**

A Final Word

The buck stops here.

Attributed to Harry S. Truman as cited by Mathews (1951, pp. 198–199).

The reader will notice that many principles enumerated in this book will require a departure from current business beliefs and practices in many organizations. Once again, these new principles are not intended to constitute a utopian or unrealistic vision. Rather, they are presented as necessary aspects of system resilience, supported by extensive research and analysis. For example, Chapter 5, Culture, lists many ways to achieve these cultural changes, both for management and for the entire organization. The fact that traditional methods of achieving these cultural changes are, on the whole, unsatisfactory, is self-evident in view of the many system failures over the past several decades.

The primary departure is the view of the enterprise infrastructure as the primary system of interest, and therefore this system must work as a whole to achieve resilience. The characteristics of this infrastructure are summarized in Chapter 7, Infrastructure. Conventional thinking does not allow for the infrastructure system to be "engineered" as product systems, that is, as hardware and software. But when the system is a collection of government agencies, for example, it is necessary to expand the definition of engineering.

The implementation aspect of this principle is that the owners of each element of the infrastructure (developers, operators, maintenance personnel, etc.) should collaborate closely to address issues of mutual responsibility. A modification of contractual processes will enhance resilience.

An important corollary to the infrastructure principle is that vulnerabilities to system resilience are not limited to design processes and that the conventional idea of "design flaws" should be viewed in a larger systemic context.

Architecting Resilient Systems: Accident Avoidance and Survival and Recovery from Disruptions.
By Scott Jackson

Any human processes, which include managerial and operational processes, may also be sources of vulnerabilities. The implication of this corollary is that review processes should be in place to assure that all human processes are performed to maximize resilience. And when they fail to perform as expected, the system will survive to perform a useful function.

A difficult question broached in Chapter 2, System Resilience and Related Concepts, is whether the principles articulated in this book would apply equally as well in the third world as in the first world. The conclusion is they are valid in either environment, but that the resources to implement them are far different.

This book has attempted to delve deeply into the implementation aspects of system resilience. Whereas many experts agree on the root causes of system failure, the question of how to implement an infrastructure is central to creating a system resilient to failure. This book, first, captures the attributes of such an infrastructure and, second, presents some actual organizational and functional options for achieving these attributes. In the end, there are two choices: accept the high level of system failures that have been observed over the past few decades or create an infrastructure that will be resilient to them. Implementation will not be easy; it involves implementing organizational and functional changes that may depart from past practices, but it should be done.

The study of resilience is in a seminal stage. The observations presented in this book represent a broad survey of the art as it stands at this time. These observations are based on the thinking of experts at this time. However, since resilience relies, to a great extent, on experience rather than on quantitative analysis, the effectiveness of the methods suggested herein remains to be validated by future generations of analysts. Perhaps more quantitative methods will emerge in the future. These topics are for future research and analysis. Perhaps there are heuristics not mentioned here. Perhaps some heuristics mentioned in this book will be superseded by others. In short, this is only the beginning of the journey, not the end.

14.1 THE CHALLENGES

Few problems are more difficult to solve than resilience. Can multiple disciplines as widely diverse as engineering and psychology, for example, be corralled to analyze resilience in its entirety? Can multiple organizations, including, for example, aircraft developers, customers, government agencies, operators, and maintenance personnel, be integrated to solve communications and decision-making problems? Can emergent events, such as conflicts between operators and software, be predicted and eliminated? Can adaptive systems be created to survive major disruptions, such as terrorist attacks, hurricanes, and human-made system failures? Can cultural change actually be achieved that will result in risk-conscious organizations that will anticipate and address threats to resilience? Finally, are there indicators of potential weaknesses in systems that can be exploited to make more resilient systems? If even one of these challenges can be achieved, then a small step will have been achieved toward reducing the likelihood of having to watch our fellow humans come to painful ends.

In addition to all the above, there is a profusion of terminology and inadequate definitions. Although human models exist, much still has to be done to meet the needs of resilience. Finally, systems engineers and other practitioners need to engage in a dialog with each other and meet the multidisciplinary challenges discussed here.

14.2 FURTHER EXPLORATION

What is your opinion? Can system resilience be implemented, or are the obstacles too great? In your opinion, what are the most profitable areas of research?

Appendix A

Domain-Specific Example for Architecting a Fire-Protection Infrastructure System

Once heuristics have been identified, the next step is to apply them to a specific domain. The purpose of this appendix is to show possible design characteristics that can be derived from the heuristics of Chapter 8, Resilience Architecting.

There is no implication that all design characteristics have been implemented in any specific fire department or that they would need to be implemented in all fire departments, because the demands of all fire departments are different. They are simply *possible* design characteristics that could be implemented.

Factors that may influence the selection of specific types of design characteristics include the following.

- Type of disruptions unique to the system, e.g., earthquakes, hurricanes, tornadoes, tsunamis, grass or forest fires, residential fires, industrial fires, and so on. Although terrorist attacks are possible in any environment, it is possible that the urban environment is more likely and would require additional design characteristics.
- Urban versus rural environment. Other geographical features may influence the decision, e.g., oceans, bays, rivers, mountains, and so on.

Architecting Resilient Systems: Accident Avoidance and Survival and Recovery from Disruptions.
By Scott Jackson

- Cost analysis may show that some design characteristics are more cost effective than others.

This appendix does not address all the heuristics of Chapter 8 because they may all apply to the fire-protection domain. For example, many heuristics apply primarily to technological systems. Although there are automated systems within the fire-protection system, most heuristics in this appendix will focus on the human and organizational aspects.

Some material in this appendix reflects a discussion on February 4, 2009 at a workshop in San Francisco organized by the Resilient Systems Working Group (RSWG) of the International Council in Systems Engineering (INCOSE) and with participation by the Anti-Terrorism International Working Group (ATIWG), the Infrastructure Working Group, and members of the San Francisco Fire Department (SFFD). Maxwell (2009) summarizes some of the main points in this meeting. Although the resilience solutions implemented by this fire department do not reflect all fire departments, they do provide a point of departure.

TYPE OF SYSTEM

When operating in a stand-alone mode, a fire-protection system is a human-intensive system. It consists of all the organizational elements of any public agency. These include the first-line firefighters, supervisors, and managers.

The term *fire protection* is used in the broadest sense because many fire-protection infrastructure systems perform missions other than fighting fires. Most fire agencies also contain paramedic personnel and equipment. They may also play a role in antiterrorism and other missions, such as earthquake and hurricane recovery. The personnel and resources to perform these missions are also elements of the fire-protection infrastructure system. The California Fire Fighter Joint Apprenticeship Committee (2002) manual contains a list of terrorism scenarios and the plan for responding to them.

The system also contains considerable equipment, which includes hardware and software. There are buildings—commonly called firehouses—trucks, ambulances, and various other vehicles. Of course, there are hoses and fire hydrants. The San Francisco Fire Department (1983) standard practices manual contains comprehensive lists of this equipment.

A key element is the communications system. This system allows communication among fire trucks, individual firefighters, firehouses, and exterior agencies, such as the police and state and federal agencies.

A fire-protection infrastructure system can also be considered an element in a larger system of systems. The larger system consists of other local, state, and federal agencies. The subject fire-protection system may be called on to assist those other agencies in fighting fires outside their own defined territory, and *vice versa*.

INTERFACES

A fire-protection infrastructure system has numerous interfaces both internal to its own system and with other systems within the larger system of systems. Interfaces include the following:

- Fuel for vehicles
- Electrical power
- Communications with internal and external elements
- Water and other fire retardants
- Medical interfaces with hospitals and other entities
- Mechanical and electrical interfaces internally and with external systems when deployed remotely

ORGANIZATION AND USE OF THIS APPENDIX

The organization of this appendix is hierarchical. This organization allows the resilience attributes to flow down to the fire-protection infrastructure system design characteristics at the lowest level. The design characteristics entries will not, of course, specify a specific solution but rather provide a list of alternatives that can be considered when making a decision.

In short, this appendix is not intended to provide a design solution to a specific fire-protection infrastructure system. Each fire-protection infrastructure system will have its own geographical, economic and political environment and types of disruptions. But rather it is intended to provide a set of considerations to keep in mind when designing such a system.

Chapter 8, Resilience Architecting, is organized according to four basic considerations: adaptability, risk, resilience prediction, and culture. This appendix will only consider adaptability. According to Chapter 8, the four attributes of adaptability are capacity, flexibility, tolerance, and interelement collaboration. This appendix will treat these attributes at the top level of the hierarchy. The hierarchy is as follows:

- Level 1—Adaptability attribute
- Level 2—Heuristic
- Level 3—Capabilities of interest
- Level 4—Options for fire protection system design characteristics

The format allows for discussion at all levels of the hierarchy. The last two levels are presented in the form of a short table. A discussion of capabilities and options is also provided.

Any given design option may involve multiple heuristics. For example, the *knowledge between nodes* heuristic requires that all elements of the fire protection infrastructure system be aware of what the other elements are doing and what their needs are. At the same time, the *functional redundancy* heuristic requires that there be multiple ways to get this information.

THE CAPACITY ATTRIBUTE

Two heuristics are of importance in determining the capacity of a fire-protection infrastructure system, as follows:

The *absorption* heuristic—The system shall be capable of absorbing a disruption.

The *margin* heuristic—The system should have adequate margin to absorb disruptions.

The *context spanning* heuristic—the system should be designed to both the worst-case and likely scenarios.

The options in Table A.1 all indicate that margin characteristics should be based on worst-case and most likely scenarios. It is recognized, first, that even known worst-case scenarios may be outside the resources of any fire-protection system. Second, unpredicted disruptions, such as terrorist attacks, may exhaust even the most robust system. Hence, that is why it is necessary to invoke the complementarity rule, as discussed in Chapter 8, and take advantage of other heuristics to deal with the disruption.

The lack of water is a persistent problem in fire protection. Additional water can be obtained in several ways, for example:

- Particularly in urban areas, underground cisterns are often installed to provide extra water.
- If a hydrant is inoperative, then lines can be extended from remote hydrants to tap extra water. This method is limited by the pressure drop in the lines.

According to Maxwell (2009), the worst case scenario chosen by the SFFD is a fire in the tunnel for the Bay Area Rapid Transit (BART) train that goes under the San Francisco Bay. The department also has a plan for a tsunami strike. As mentioned in Chapter 8, Resilience Architecting, preparing for this

Table A.1. Fire-Prevention Capabilities and Options for Capacity

Capabilities	Options
Fire suppressant	Sufficient water for worst-case and most likely scenarios. This may include complete loss of water supply.
	Sufficient foam for worst-case and most likely scenarios
Modes of delivery	Trucks, helicopters, fixed wing aircraft, all-terrain vehicles, and bulldozers.
Personnel	Sufficient firefighters for worst-case scenario.

worst-case scenario also provides resilience to other possible disruptions, such as a terrorist attack on the tunnel.

In addition to the main water supply, the SFFD relies on underground cisterns as a backup to the main system.

The *functional redundancy* heuristic—An alternative method should exist to perform each critical function that does not rely on the same physical systems. Table A.2 provides examples of how functional redundancy can be achieved.

THE FLEXIBILITY ATTRIBUTE

The *reorganization* heuristic—The system should be able to restructure itself in response to disruptions or the anticipation of disruptions. Table A.3 provides a list of options for reorganization.

Table A.2. Fire-Prevention Capabilities and Options for Functional Redundancy

Capabilities	Options
Fire suppression	There should exist multiple ways to suppress fires.
Modes of delivery to disruption	There should exist multiple modes of transportation to travel to the disruption.
Routes to disruption	There should exist multiple routes to the disruption
Personnel	Nonregular personnel should be available and properly trained, e.g., personnel from remote locations and/or volunteers.
Vertical movement	Various methods should be available to reach upper floors of tall buildings: ladders, helicopters, ropes, and so on.
Operations on water	Various methods of performing duties on water: boats, swimming, diving, aerial, and so on.
Fire protection skills	Firefighters should be crosstrained in all major skills.
Communications redundancy	There should be multiple ways to communicate with all internal and external nodes of the fire-protection infrastructure.

Table A.3. Fire-Prevention Capabilities and Options for Reorganization

Capabilities	Options
Capability to reorganize in the face of different disruptions	Standard procedure to authorize new ad hoc organizations as needed within the infrastructure system.
	Standard procedure to empower lower level organizations within the infrastructure system.
Capability to perform mission outside the boundary of stand-alone system	Training, equipment, personnel for extra-system operations.

According to Maxwell (2009), the SFFD puts much emphasis on fire code and earthquake code update. A code update would, of course, apply to any fire department. Of course in other areas, tornado codes and hurricane codes would be the area of focus. The SFFD has asked for help in developing training scenarios for other disaster scenarios, such as terrorist attacks.

Maxwell also notes how the SFFD has a three-level decision-making structure. Firefighters are trained to be aggressive, mid-level officers are trained to be aggressive and also logical, and incident commanders are trained to be conservative to avoid conflicts or confusion during an incident.

The *diversity* heuristic—There should be diversity within systems. Table A.4 provides options for diversity in a fire-protection infrastructure.

THE TOLERANCE ATTRIBUTE

The *graceful degradation* heuristic—The system should degrade gradually when exposed to a disruption. Table A.5 provides options for graceful degradation in a fire-protection infrastructure.

The SFFD relies on a water system with graceful degradation. Shut-off valves are positioned at various points in the system so that if the water system fails at any point, this section of the system can be shut off, and the remainder of the system will continue to function, that is, supply water.

The *drift correction* heuristic—Drift toward brittleness should be detected and corrected. Table A.6 provides options for drift correction in a fire-protection infrastructure.

The *neutral state* heuristic—The system should be put into neutral if possible. Table A.7 provides options for achieving neutral states in a fire-protection infrastructure.

Table A.4. Fire-Prevention Capabilities and Options for Diversity

Capabilities	Options
Capability to interface with diverse external systems	Adapters for all external systems, electrical, mechanical, and information interfaces Standardization of interfaces through external agreements

Table A.5. Fire-Prevention Capabilities and Options for Graceful Degradation

Capabilities	Options
Avoidance of resource depletion	Standard procedure and training to avoid concentration of resources in the face of a disruption. Making elements of the system not tightly coupled, that is, vulnerable to a series failure.

The SFFD implements the *neutral state* heuristic by training the firefighters to secure the area in the case of an emergency disruption.

The *organizational decision making* heuristic—Organizational decision making should be monitored.

The *organizational planning* heuristic—Notice signs that call into question organizational plans, models, and routines.

Many references, for example, Reason (1997) and Leveson (1995), note that optimum decision making is achieved when decisions are delegated to the lowest level of the organizational infrastructure. Of course, policies and training need to be in place to make this happen. Table A.8 provides a list of options for achieving decision making and organizational planning in a fire-protection infrastructure.

Table A.6. Fire-Prevention Capabilities and Options for Drift Correction

Capabilities	Options
Gain knowledge of disruption before it happens or as soon as possible after it happens	Immediate access to advance warning systems (see inter-element collaboration)
React to disruption before it happens or immediately after	Ready plans and training for all anticipated disruptions or for disruptions of an unpredicted nature

Table A.7. Fire-Prevention Capabilities and Options for Achieving a Neutral State

Capabilities	Options
Achieve a neutral state	Muster points
	Standard procedures and training for action immediate after a disruption
	Fire protection personnel should focus on securing the area and saving lives as a first priority

Table A.8. Fire-Prevention Capabilities and Options for Decision Making and Organizational Planning

Capabilities	Options
Assure optimum decision making	Standard policy for decision-making empowerment at the lowest level
	Training on decision making
	Review of decision making in case studies and incorporation into procedures and policies

The mobility heuristic – The system should be able to avoid a threat by moving. Table A.9 provides options for achieving mobility in a fire protection infrastructure.

THE INTERELEMENT COLLABORATION ATTRIBUTE

The following heuristics apply:

The *knowledge between nodes* heuristic—Maximize knowledge between nodes.

The following capability of communication among nodes, not just within the fire-protection system but also with other fire-protection systems, requires that the communication systems be interoperable, that is, their communications equipment is compatible. In past disaster, for example Hurricane Katrina, it was found that many local communication systems were not interoperable with each other, according to Stephan (2007). Table A.10 provides a list of options for achieving knowledge between nodes in a fire-protection infrastructure.

The *intent awareness* heuristic—Each element of the system should have knowledge of the others' intent and should back up each other when called on.

Table A.9. Fire Prevention Capabilities and Options for Mobility

Capabilities	Options
Movement	Various methods should be employed to move in the event of a threatening disruption: trucks, cars, motorcycles, bicycles, foot, and so on.
Communications	Communications nodes, especially command posts, should be mobile.

Table A.10. Fire-Prevention Capabilities and Options for Knowledge Between Nodes

Capabilities	Options
Disseminate knowledge	Knowledge among personal elements: telephone, and radio Knowledge from command structure: telephone, and radio Knowledge from exterior nodes: telephone, and radio

Table A.11. Fire-Prevention Capabilities and Options for Achieving knowledge of Other Nodes' Intent

Capabilities	Options
Intent	Standard protocol in discussions with nodes both within and without the fire-protection infrastructure system

Table A.12. Fire-Prevention Capabilities and Options for Avoiding Impediments to Inter element Collaboration

Capabilities	Options
Eliminate procedural impediments	Advance agreements with external agencies, such as federal, state, county, other fire-protection agencies, and other agencies.
Eliminate operability impediments	Interoperable communications with all external nodes
Eliminate communications clogging	Communications architecture to avoid clogging

Table A.11 provides a list of options for achieving knowledge of other nodes' intent.

The *interelement impediment* heuristic—There should be no impediments to interelement collaborations.

These agreements should be detailed with regard to resources, timing and other aspects. If the higher level agencies do not take the lead on this capability, then the fire-protection infrastructure management should do so. Table A.12 provides a list of options for avoiding impediments to interelement collaboration.

The *interelement collaboration* heuristics are among the most important in the fire protection domain.

Appendix B

A Resilience Analysis of Metrolink 111

INTRODUCTION

Chapter 4, Case Histories, provides a brief summary of the Metrolink 111 accident on September 12, 2008, in Chatsworth, California, near Los Angeles. This appendix provides a deeper look at the accident from a resilience point of view. It asks key questions, such as, what was the nature of the disruption? What were the brittleness aspects, that is, lack of resilience, of the system in which the commuter train operated? What are the options for increasing the resilience of the system? Who is responsible for implementing these measures? Although the cost of implementing these measures cannot be estimated at this time, a qualitative look at the costs can be made.

This case provides a comprehensive view of the complexity of a transportation infrastructure, the many stakeholders involved, and the difficulty of creating a resilient system of this type. The train involved in the accident was similar to the one shown in Figure B.1.

OVERVIEW OF THE ACCIDENT

On September 12, 2008, a Metrolink commuter train collided with a Union Pacific freight train killing 25 people and injuring many more in Chatsworth, California, near Los Angeles. What is not disputed is that the Metrolink train did not stop at a signal that would have prevented it from the collision. Initial

Architecting Resilient Systems: Accident Avoidance and Survival and Recovery from Disruptions.
By Scott Jackson

Figure B.1. A Metrolink train.
Photo taken by author.

attention was focused in the train's engineer, Mr. Robert M. Sanchez, who, according to Lopez and Hymon (2008a), was thought to have been using his cell phone to send text messages at the time of the crash. According to Archibold (2008), the National Transportation Safety Board (NTSB) determined that Sanchez did not apply the train's brakes before the collision. Metrolink attempted to contact Sanchez shortly before the accident, but there was not enough time.

Some witnesses have disputed this account, but that aspect will be discussed below. Also not in dispute is the fact that Metrolink and Union Pacific shared a single track for a portion of their routes.

The accident was classified as a "delay in block" accident. GCOR (1994, Section 9.9) defines delay in block to be "If a train has entered a block on a proceed indication that does not require restricted speed, and the train stops or its speed is reduced below 10 MPH, the train must: (A) proceed with restricted speed, (B) proceed prepared to stop at the next signal until the next signal is visible and that signal displays a proceed indication, or (C) operate according to cab signal indication."

A second Metrolink accident occurred in Rialto, California, only 2 months after the Chatsworth accident. Rialto is also near Los Angeles. The Rialto accident bore similarities to the Chatsworth accident in that the Metrolink train ran a red light. One difference was that the train in Rialto had two engineers rather than one.

Prior to the Chatsworth (Metrolink 111) and Rialto accidents, there had been three other Metrolink accidents according to Rohrlich (2008). The first was in Placentia, California, on April 23, 2002, with 3 fatalities and 162 injuries. The second was in Burbank, California, on January 6, 2003, with 1 fatality and 30 injuries. The third was in Glendale, California, on January 26, 2005 with 11 fatalities and 180 injuries.

THE DISRUPTION

Chapter 3, Disruptions, provides two categories of disruption, Type A and Type B. Type A disruptions are externally caused, such as terrorist attacks or natural forces like as hurricanes. Type B disruptions are internally caused, such as software or human errors. These disruptions are called systemic disruptions because the causes are internal to the railway transportation system. The Metrolink 111 accident was a Type B systemic disruption.

The Human Error Theory

From the outset, the primary disruption theory has been that the train's engineer, Robert M. Sanchez, was using his cell phone immediately before the accident. According to Lopez and Hymon (2008a), the NTSB has shown that Sanchez had sent a message only 22 seconds before the impact. According to Morrison (2008), Metrolink itself immediately issued a statement that the engineer was responsible. Sanchez himself was not an employee of Metrolink but of another company who provides services, such as those provided by Sanchez, to Metrolink. Metrolink later filed a lawsuit against that company.

Another factor considered to be a contributor to the accident was the split-shift work hours. With this system, the operators would work 5 hours, take a break, and then work another 5 hours. According to Lopez and Hymon (2008b), federal sources report that tired work crews have been a cause of some of the deadliest accidents in recent history.

The Faulty Signal Theory

Weeks after the accident, an alternative disruption theory emerged. According to Hennessey-Fiske et al. (2008), three witnesses testified that the signal the engineer was supposed to see was green rather than red. If this was the case, then the engineer may have seen the green signal and proceeded, irrespective of the cell phone issue. According to Lopez and Connell (2008b), the statements by the three witnesses were later corroborated by the only surviving crewman, Robert Heldenbrand, on the train. Heldenbrand also argues that the fact that the crew did not confirm the color of the signals as they passed them was because, according to Heldenbrand, the signals were green. If the signals had been red or yellow, the crew would have been required to call out the signals. This latter argument adds some credibility to the faulty signal theory.

However, according to Lopez and Oldham (2008), the NTSB had previously released a statement that the signal was working properly, thus contradicting the witnesses. The NTSB has stood by its previous statement.

The Signal Gap Theory

Three months after the accident, according to Lopez and Connell (2008a), a group of safety experts not related to the NTSB posed another theory, namely, that the fact that there was a station between the two signals may have created a distraction for the engineer and caused the disruption. Although this theory may be part of the NTSB's final report, they have not commented on it at the time to the independent safety group statement.

Conclusions Regarding Disruptions

Gottlieb (2008d) states that the NTSB's preliminary finding was that the cause of the accident was "human error." Of course, one of the themes of this book is that there is a distinction between a disruption and an accident. Although the actions of the engineer may have been ill-advised and even illegal, and such actions cannot be excused, the conclusion is that the real cause of the accident was the brittleness, or lack of resilience, of the system. What actions should be taken to make the system resilient? This question will be addressed in later paragraphs.

In a discussion of the Metrolink 111 accident, Meshkati and Osborn (2008, p. A23) make the following statement regarding the human error theory:

> The root cause of most of these [Metrolink] accidents is attributed to human error . . . Flatly attributing accidents to just the operators—or drivers—is an oversimplification of the problem.

Then there is the faulty signal theory. So there are many possibilities. One is that the engineer was indeed the cause of the disruption, or the faulty signal may have been green, or both may have been true. The latter scenario would fit the model that is so often true for catastrophes, namely, that accidents result from multiple causes. Given the uncertainty in the cause of the disruption, it would be more prudent to assume that both are true so that any approach to resilience would cover both cases. It will be dicussed later in this appendix that some of the suggested approaches only address the human error cause.

It is not the purpose of this appendix to resolve the disruption theory because it does not really matter which theory is correct. The important thing is that disruptions will happen by many causes, and these are only two possibilities. The disruption, or disruptions, occurred within a brittle system.

It is the brittle system that should be addressed. This brittleness will be addressed in the following paragraphs.

Once again, the possibility of human error, such as sending text messages when vigilance is required or possible faulty signals, should not be ignored. They are only elements within a larger system that should shed its brittleness and become resilient.

THE SYSTEM

As discussed in Chapter 2, System Resilience and Related Concepts, when addressing the resilience of a system, one should define what system it is that needs to be analyzed. The system of interest in the Metrolink 111 accident is a composite of multiple interlocking systems. That is, it is a system of systems. Figure B.2 provides a sketch of the systems involved in this system of systems. Following is a discussion of these systems. For simplification, all possible links between system elements are not shown.

Metrolink

Metrolink is the operating name of the Southern California Regional Rail Authority (SCRRA) headquartered in Los Angeles. Metrolink was set up by the Joint Powers Authority (JPA) of the counties of Los Angeles, Orange, San Bernardino, Riverside, and Ventura. The Metrolink system consists of all the trains, personnel, and support equipment required to operate and maintain the Metrolink trains. Metrolink itself has already decided to mandate two engineers in a train cab on selected routes. Solutions beyond this one would probably be beyond Metrolink's capability. Metrolink has also already forbidden the use of cell phones by operators on duty. Also, after the accident, according to Gottlieb (2008c), Metrolink established an advisory safety panel made of safety experts. The role of the panel is to "take a look at the commuter rail's safety and operating procedures."

Metrolink owns track and other property purchased in fee title from Burlington Northern Santa Fe (BNSF) railways and railroad Union Pacific (UP).

Metropolitan Transportation Authority (MTA)

The MTA is a local government agency that oversees and provides a portion of the funding for the Metrolink system.

Veolia Transportation

Veolia Transportation is a private company whose subsidiary employs the engineers on the Metrolink trains.

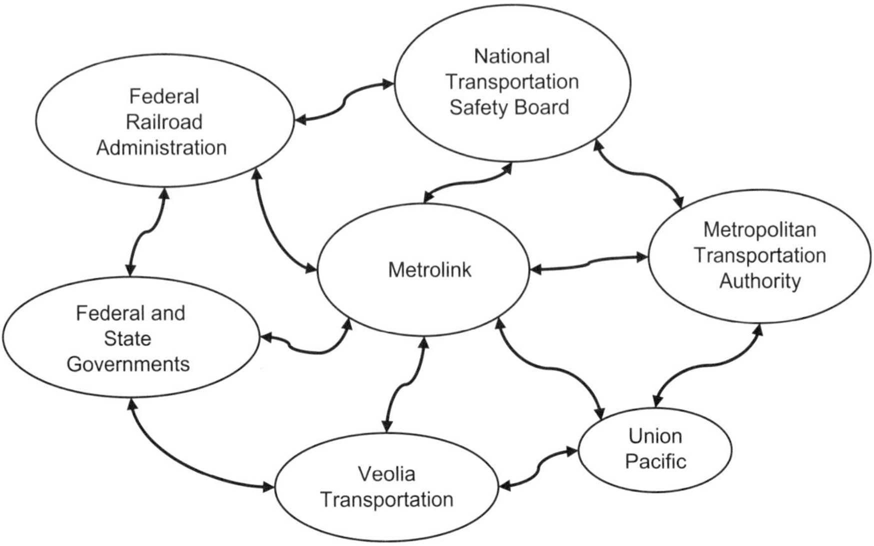

Figure B.2. The Metrolink system of systems.

Union Pacific

The Union Pacific is a railroad company engaged in the transfer of cargo by means of freight trains. The Union Pacific also owns the tracks on which the Metrolink line was operating at the time of the accident. The Union Pacific owns the train that collided with the Metrolink train. According to Rohrlich (2008), both Union Pacific and Burlington Northern Santa Fe railway companies had resisted automated train safety devices as too expensive.

Federal and State Governments

This element of the Metrolink system of system consists primarily of the U.S. government and the California state government. These governmental entities play two important roles in the Metrolink system of systems.

The first role is funding. Some suggested solutions to achieve resilience are costly. For example, the funding to end single-track operations would most likely have to come from federal sources. This step would consist of constructing parallel tracks where single tracks now exist. Funding to provide the advanced warning systems would also most like come from the federal government.

The second role is legislative action. We have already discussed legislation by the California government regarding the use of cell phones on trains. In addition, the use of the advanced warning system on a national level would require a federal mandate.

Meshkati and Osborn (2008, p. A23) have recommended a commission at the state level in California "to address the fundamental problems of rail safety."

One state agency, the California Public Utilities Commission, has already taken one step in response to the Metrolink 111 accident. According to Connell and Lopez (2008b) and Hymon (2008a, October 3), this commission has banned the use of cell phones by all train crews. Although this ban addresses only the human error disruption issue, it is one step towards addressing rail safety. It also illustrates the multiplicity of responsible government agencies that can play a role in the rail safety problem.

National Transportation Safety Board (NTSB)

The NTSB is an independent government board NTSB established by congress with the purpose of investigating accidents. The NTSB normally makes recommendations with respect to safety; however, the NTSB does not have the authority to ensure that the recommendations are implemented.

Federal Railroad Administration (FRA)

The FRA is an administration within the Department of Transportation charged with establishing safety standards for railroads.

BRITTLENESS OF THE SYSTEM

In this section, we will discuss the many ways in which the Metrolink system is brittle, that is, lacks resilience. In the following section, we will discuss the various ways that have been proposed to improve resilience. This discussion will be within the framework of the four attributes of resilience as discussed in Chapter 8, Resilience Architecting.

Capacity

The basic capacity deficiency of the Metrolink system is that it has no way to absorb a disruption of the scale of the Metrolink 111 accident. Of course, designing trains to collide with one another is out of the question, so the most important capability is functional redundancy, that is, the ability to achieve a given function in multiple ways. The single-track system is the primary impediment to functional redundancy.

However, the primary deficiency in functional redundancy is the lack of multiple ways to detect an impending collision. The dependency on a single engineer and a single signal to accomplish this function is the primary deficiency in functional redundancy. In safety terminology, this situation is known as a single-point failure.

In the context of the Swiss cheese model discussed in Chapter 3, Disruptions, in which it is assumed that all high-consequence systems have at least two layers of defense, it can be concluded that Metrolink has only one layer of defense. Because the signal and the operator are so interdependent, they cannot be considered different layers. This single layer of defense contributes to the system's brittleness.

Flexibility

The primary flexibility guideline is the *reorganization* heuristic. The *reorganization* heuristic is manifested through various means to avoid catastrophe. The Metrolink system had a single means, namely, the dependence on a single engineer to take appropriate action. Alternative means are discussed in paragraphs to come.

Tolerance

The operating heuristic for the tolerance attribute is the *drift correction* heuristic. This heuristic requires the knowledge that a failure is imminent. Hence, advance notice is paramount. In the Metrolink, case there did not seem to be any advance notice at all. Even if the engineer had been totally alert, the time required to avert disaster was on the order of seconds, not minutes. Hence, it can be said that the safety of the Metrolink system was dangerously close to a perilous cliff. In the paragraphs to follow, various methods to improve the anticipation factor are summarized.

Interelement Collaboration

System resilience requires communication and collaboration among all key elements of the system. For example, was there any communication between the engineer and any other element of the system? Apparently there was not. Was there any communication between Metrolink and the Union Pacific freight train? Apparently there was not. The communication between the signal and the engineer apparently failed for one of the two disruption theories discussed above. Many other heuristics were not invoked, including the *knowledge between nodes* heuristics. So, in short, the engineer was an element of the system acting in isolation. The net result was a fragile system.

Other evidence of the interelement brittleness of the Metrolink system appeared with the second Metrolink accident in Rialto, California, on November 20, 2008, only 2 months after the Chatsworth accident.

Also, as pointed out above, both the Union Pacific and Burlington Northern railway companies had resisted automated safety devices as being too expensive. Thus, it can be concluded that far from collaboration among the nodes of the Metrolink system of systems, there is evidence of resistance to collaboration.

Chapter 8, Resilience Architecting, points to the *interelement impediment* heuristic, namely, that there should be no impediments to interelement collaborations. Metrolink is an example of impediments to interelement collaboration contributing to its brittleness. According to Rohrlich (2008), agreements between Metrolink and the freight train lines do not give Metrolink the power to obligate the freight train companies to safety devices.

Cultural Aspects

Throughout the reporting on the Metrolink 111 crash, various cultural aspects have come to light that reflect the paradigms discussed in Chapter 5, Culture. Probably the most salient paradigm at work on Metrolink was the conflicting priorities paradigm, which was discussed in Chapter 5. In short, the conflicting priorities were cost versus safety, as discussed by Rohrlich (2008). Although Metrolink had many opportunities to make the case for more advanced safety equipment, they did not. The head of Metrolink, Mr. David Solow, has come under considerable criticism for his failure to emphasize safety, as discussed by Gottlieb (2008a).

Another cultural aspect is the possibility of an ethics lapse on the part of the firm that supplies Metrolink with engineers. According to Lopez and Hymon (2009), an employee had testified that she had complained before the accident about text messaging by the engineer.

RESILIENCE ENHANCEMENTS

Various approaches have been suggested for enhancing the resilience of the Metrolink system. Some of these have actually been adopted and will be implemented. This section will discuss these enhancements and the degree to which they will improve resilience.

The Parallel Track Method

The placing of a second track parallel to the existing track is the most secure, and probably the most costly, resilience method. With this method the likelihood of a collision, by whatever disruption, is highly unlikely. This method relies entirely on achieving the attribute of capacity, as described in Chapter 8, Resilience Architecting. The heuristic of importance is functional redundancy. That is, with two methods of performing the transit function, the failure of one method will not affect the other.

The principal difficulty with this method is its cost, not only the cost of laying extra track but also the cost of acquiring additional land. For these reasons, it is unlikely that this solution will be effected in a short time or by local entities, such as the Metropolitan Transportation Authority (MTA). It is acknowledged that single-track systems are uncommon on the west coast of the

United States and more common on the east coast. The most likely source of funds for a parallel track system is the U.S. federal government.

The Second-Engineer Method

One of the first methods discussed is the idea of having a second engineer in the cab of the Metrolink train to assist in observing signals. Although this method has its limitations, its chief advantage is that it was possible to implement immediately after the accident. According to Hymon and Connell (2008), the second-engineer method is one of three approaches being considered by the MTA. In fact, according to Gottlieb (2008b), Metrolink management announced 15 days after the accident that it was implementing this method.

The primary heuristic satisfied by the second-engineer method is the *functional redundancy* heuristic. That is, if one engineer does not see the signal, the second engineer may see it, thus increasing the probability that the signal will not be missed. This method, however, does not assure that the signal will always be seen.

The second-engineer method also enables the capability of the system to satisfy the *drift correction* heuristic. That is, with more advance notice of an impending failure, there is an increased ability to correct the problem.

A shortcoming of the second-engineer method is that it only provides protection against one type of disruption, namely, the failure of the first engineer to see the signal. If the faulty signal was the cause of the disruption, as described above, then the second engineer would provide no protection at all against this type of disruption.

Another shortcoming of this method is that there is no guarantee that the second engineer will be more diligent than the first one. Despite all these shortcomings, immediate implementation will provide some measure of resilience not present in the system before the accident.

As we saw before, two engineers in the cab did not seem to provide much benefit in the Rialto accident.

Filming of Crew

According to Connell (2008), Metrolink has announced its intention to film the crew of the Metrolink trains during operation. The purpose of this method is to know exactly what is happening in the cab. This proposal has received much criticism. The principal objection is that the on-board cameras would do nothing to prevent future crashes but rather to know after a crash what has happened. It could be concluded that that another limitation of this method is that it does nothing to improve the resilience of the entire Metrolink system. It only addresses one type of disruption.

Positive Train Control

Many sources have described an advanced detection system that is currently being used in other parts of the United States but not in California. This system uses Global Positioning System (GPS) satellite tracking data to determine exactly where every train is and how close they are to each other. With this system, a signal can be sent to any train to cause it to stop if another train is on a collision path. According to Lopez and Weikel (2008), his method is sometimes called "positive train control."

This method would satisfy both the functional redundancy and the *drift correction* heuristic. So, it would not matter which type of disruption occurs, either the human error or the faulty signal disruption. This system would stop the train regardless of the disruption. According to Hymon and Connell (2008), positive train control is one of three approaches being considered by the MTA.

However, care would have to be taken to assure that the advanced detection system itself cannot be the source of a single-point failure. This assurance can be gained by employing the *human backup* heuristic, as discussed in Chapter 8, Resilience Architecting. Hence, the use of the advanced detection system and a human backup, such as the engineer, would yield increased resilience.

Although the cost of the advanced detection system has not been widely published, its cost has been stated to be high. Therefore, its implementation would need to rely on higher level funding sources, such as the federal government. Lopez and Weikel (2008) state that cost was the main reason that Metrolink has resisted the need to implement such a system. This statement was based on the testimony of California's Senator Feinstein. Senators Feinstein and Boxer of California have introduced legislation to make this system mandatory. According to Lopez and Hymon (2008a), the U.S. Senate has overwhelmingly approved this legislation.

Another factor is the time of development. It is generally agreed that the implementation of the advanced detection system would take some time, in any event, longer than other methods. Therefore, resilience could be increased by using other methods while the advanced detection system is being implemented and funding is being sought.

Finally, one of the strongest deterrents to positive train control is the interjurisdictional aspect because this system would have to be adopted by Metrolink, Union Pacific, and Burlington Northern Santa Fe railroads. Thus, the deployment of such a system would require the action of a higher level authority, such as the NTSB or other government entity.

Automatic Train Stop

Many sources have discussed the use of an existing technology known as the automatic train stop. The principal advantages of this technology are its lower cost and faster implementation. According to Lopez and Weikel (2008), the existing technology works by sounding an alarm if the train does not respond to a signal. If the engineer does not heed the alarm, the system will stop the train automatically, which may be after the train has passed the signal. Thus, the train may stop too late. However, most experts agree that its performance is not as good as the advanced detection system discussed above. Instead of stopping the train, the existing technology might only slow the train. According to Rohrlich (2008), this method would probably not have prevented the Metrolink 111 crash.

The automatic train stop system is dependent on rules that the operator is required to follow. If the train passes a signal that is anything less than clear, the engineer must "forestall," that is, he must acknowledge the signal. If he does not, the train automatically stops.

However, the fact that performance is not high should not rule the existing technology out altogether. The existing technology, when used in combination with other methods, may provide a resilient system and be cheaper and faster to implement than other methods. According to Hymon (2008b), Metrolink has agreed to consider the installation of the automatic train stop system.

Cab Signaling

Another system used by some railroads is the cab signaling system, which is also called the visual cue system. This system receives real-time sensor location and speed data from the track that is displayed in the cab of the train. This information can be used either to notify the engineer or to stop the train automatically. According to Rohrlich (2008), this is an example of an old technology that would have prevented the Metrolink 111 accident.

Video Cameras

Another feature that has been discussed is the addition of video cameras on the trains. The primary advantage of these cameras is that they would assist the human operators in their job of detecting either the signal or the oncoming trains. Another advantage is that these cameras can be installed quickly and at moderate cost.

The chief disadvantage is that they still rely on the observation of humans, that is, the engineer or engineers. Hence, these cameras would only increase the observation capability of the humans.

According to Hymon and Connell (2008), the installation of video cameras is one of three approaches being considered by the MTA.

Another disadvantage is that they only address one type of disruption, that is, the human error disruptions. In the Metrolink 111 case, if the signal was indeed faulty, the cameras would do little to detect the oncoming train unless the closing speed is slow.

CONCLUSIONS AND APPROACHES

We saw in Chapter 10, Measuring Resilience, that although many aspects are measurable, some are not. The aspects with limited measurability include human error, unpredicted disruptions, and the effectiveness of heuristics.

With these aspects in mind, it is useful to categorize solutions in terms of short-term, medium-term, and long-term solutions. It is also useful to consider what *combinations* of solutions would make the most sense. Some of these methods are already being considered or have already been identified for implementation. However, it is not relevant what methods are actually selected; what is important is an understanding of which methods yield the most resilience for the least cost and shortest time of implementation.

Short-Term Implementation

The approaches that are feasible in the short term consist of a combination of the second-engineer method, video cameras, and the automatic train stop method. Although any one of these methods would probably not yield a highly resilient system, the combination of methods is more likely. All these methods are the most viable from a cost and schedule point of view. It is unlikely that appeals to state and federal agencies will be required.

The second step that can be made in the short term is to establish and enhance a collaborative relationship with the other members of the system shown in Figure B.2. This step would be in accordance with the *knowledge between nodes* heuristic discussed in Chapter 8, Resilience Architecting. This relationship can be established, where it may not exist already, with the MTA, the Union Pacific railroad, state national government agencies, and the National Transportation Safety Board. Although relations may already exist with these entities, Metrolink can examine these relationships to determine where they can be enhanced. This responsibility for this relationship may lie either with Metrolink management or the Metrolink safety panel discussed by Gottlieb (2008c).

On December 12, 2008, according to Connell and Lopez (2008a), Metrolink received the report of a Peer Review Panel, which made the following recommendations:

- Metrolink should increase oversight of contract workers, such as the engineer who allegedly was text messaging on his cell phone. This oversight would consist of medical and psychological testing. This step would be in agreement with the discussion on *Training for humans on the sharp edge* in Chapter 6, Capabilities.
- Metrolink should appoint an executive with an emphasis on safety. This step is in agreement with the organizational structure discussed in Chapter 7, Infrastructure, which calls for an emphasis on resilience at the top of the organizational structure.

- Metrolink should train current executives in the principles of safety. This step would address the *high-level problem paradigm* in Chapter 5, Culture, in which executives tend to dismiss safety as a low-level topic not within their sphere of interest.

The panel also recommended implementation of positive train control, which is discussed below under long-term implementation.

Medium-Term Implementation

The major addition to the above approaches in the medium term is the advanced detection system, known as "positive train control." The combination of that system with all of the short-term approaches would yield a highly resilient system. At a minimum, the use of the human backup with the advanced detection system would satisfy the *functional redundancy*, *drift correction*, and *human backup* heuristics.

A bill authorizing this system, called the Rail Safety Improvement Act, has passed through Congress and has been signed. However, according to Rohrlich (2008), it will not be mandatory until 2015, illustrating the time lag in implementing this approach. This bill authorizes $50 million for each fiscal year from 2009 through 2013.

Long-Term Implementation

As discussed above, the approach that would provide the greatest resilience is the parallel track solution. Although this solution would be most effective, the cost and time to implement it may be large. This approach would require legislative approval at the government level, which may be lengthy.

References

Anders, S., Woods, D., Wears, R., Perry, S., & Patterson, E. (2006). Limits on adaptation: Modeling resilience and brittleness in hospital emergency departments. *Proceedings of the Second Symposium on Resilience Engineering*, (pp. 1–9). 8-10 November, Antibes-Juan-les-Pins, France.

ANSI/EIA-632-1998. (1999). *Processes for engineering a system*. Philadelphia, PA: Electronic Industries Association.

Archibold, R. (2008, September 17). Commuter train did not use brakes, officials say. *New York Times*, p. A15.

Ashkenas, R., et al. (1998). *The boundaryless organization : Breaking the chains of organizational structure*. San Francisco, CA: Jossey-Bass.

Augustine, N. (1996). Yes, but will it work in theory? *1996 Woodruff distinguished lecture transcript*. Retrieved March 8, 2009, from http://sunnyday.mit.edu/16.355/Augustine.htm

Aviation Safety Network. (1989). *Accident description (1989 f Sioux City DC-10)*. Retrieved on March 8, 2009, from http://aviation-safety.net/database/record.-php?id = 19890719-1

Smith, J. (2007, March 5) *Aviation Week and Space Technology* (2007, March 5), *166*, 23.

Baker, J. III, et al. (2007). The report of the BP U.S. refineries independent safety review panel. *Oil, Gas, & Energy Law Intelligence*, 2.

Barter, R., & Morais, B. (1998). Systems engineering applied to integrated safety management for high consequence facilities.. *9th Annual International Symposium of the Council on Systems Engineering*. 6 June, 1999, Brighton, England.

Bellis, M. (2009). The history of penicillin. *About.com* : *Inventors*. Retrieved on March 8, 2009, from http://inventors.about.com/od/pstartinventions/a/Penicillin.htm

Architecting Resilient Systems: Accident Avoidance and Survival and Recovery from Disruptions.
By Scott Jackson

Bennis, W. (1999). The leadership advantage. *Leader to leader*, *12*, 18–23.

Besco, R. (1990). Aircraft accidents aren't. *Accident Prevention*. Flight Safety Foundation. 1990 (Part 1); 1991 (Part 2).

Billings, C. (1997). *Aviation automation: The search for a human-centered approach*. Mahwah, NJ: Lawrence Erlbaum Associates.

Blanchard, B., & Fabrycky, W. (2006). *Systems engineering and analysis* (4th ed.). Englewood Cliffs, NJ: Prentice Hall.

Bracco, F., Gianatti, R., & Pisano, L. (2008). Cognitive resilience in emergency room operations, a theoretical framework. *Proceedings of the Third Symposium on Resilience Engineering*. 28–30, October, Antibes - Juan-les-Pins, France.

Branson, R. (2007, July 29). Richard Branson's empire keeps growing. *CBS News Sunday Morning*. New York: Columbia Broadcasting Company.

Bureau Enquétes Accidents (BEA). (2000). Accident on 25 July 2000 at "La Patte d'Oie" in Gonesse (95), to the Concorde, registered F-BTSC, operated by Air France. Retrieved on March 27, 2009, from http://www.bea-fr.org/docspa/2000/f-sc000725pa/pdf/f-sc000725pa.pdf

California Fire Fighter Joint Apprenticeship Committee. (2002). *First responder: Operations – terrorism consequence management*. Student Manual. California Fire Fighter Joint Apprenticeship Committee: Sacramento, CA.

Carnegie Mellon Institute. (2006). Capability maturity model integration. Retrieved March 13, 2009, from http://www.sei.cmu.edu/cmmi/

Carroll, P. (1999). The executive leader's perspective. In Senge, P., et al. (eds.) *The dance of change: The challenges to sustaining momentum in learning organizations* (pp. 203–211). New York: Doubleday.

Checkland, P. (1999). *Systems thinking, systems practice*. New York: John Wiley and Sons.

Chiles, J. (2002). *Inviting disaster: Lessons from the edge of technology, an inside look at catastrophes and why they happen*. New York: Harper Business.

Columbia Accident Investigation Board. (2003). *Columbia accident investigation report*. Washington, D.C.: National Aeronautics and Space Administration.

Commercial Aviation Safety Team (CAST). (2007). Safer skies/CAST: 65 selected safety enhancements. Retrieved on March 27, 2009, from http://www.cast-safety.org/pdf/safety_enhancements_completed.pdf

Connell, R. (2008a, September 17). Metrolink seeks to film its engineers. *Los Angeles Times*, p. B1.

Connell, R., & Lopez, R. (2008b, December 5). Safety of rail system assailed. *Los Angeles Times*, pp. B1, B10.

Connell, R., & Lopez, R. (2008c, September 19). Cellphones banned for train crews. *Los Angeles Times*, p. B5.

Conrow, E. (2000). *Effective risk management: Some keys to success*. Reston, VA: American Institute of Aeronautics and Astronautics (AIAA).

Critical thinking: Moving from infrastructure protection to infrastructure resilience. (2007). CIP [Critical Infrastructure Protection] Program Discussion Paper Series. George Mason University. Retrieved on March 27, 2009, from http://www.resilient.com/download/Research_GMU.pdf

D'Antonio, M. (2002, July 14). The advantage of falling short. *Los Angeles Times*. pp. 12–15, 32–33.

Dean, C. (2006, May 2). Engineering a safer, more beautiful world, one failure at a time. *New York Times*. p. D3.

Dekker, S. (2006). Resilience engineering: Chronicling the emergence of confused consensus. In Hollnagel, E., Woods, D. D., & Leveson, N. (Eds.). *Resilience engineering: Concepts and precepts*. (pp. 77–92). Aldershot, UK: Ashgate Publishing.

Dekker, S., & Hollnagel, E. (2006). *Resilience engineering: New directions for measuring and maintaining safety in complex systems*. Second Progress Report, School of Aviation, Lund University. Retrieved on March 27, 2009, from http://www.srv.se/upload/Om%20verket/Forskning/projekt/Rapporter/Dekker%20_Progress%20Report_2.pdf

Department for Environment, Food and Rural Affairs (Defra). (2007). *Making space for water: Outcome measures*. Retrieved on March 8, 2009, from http://www.defra.gov.uk/environ/fcd/policy/strategy/sd4/default.htm

Department for Environment, Food and Rural Affairs (Defra). (2008). *Developing the evidence base for flood resilience: Technical Summary FD 2607*. Retrieved on March 27, 2009, from http://sciencesearch.defra.gov.uk/Document.aspx?Document=FD2607_7321_TSM.pdf

Department of Defense. (2001). Technology readiness levels and their definitions. In *Mandatory procedures for major defense acquisition programs (MDAPS) and major automated information system (MAIS) acquisition programs* [DoD Deskbook 5000. 2-R, Appendix 6] (pp. 183–185). Washington, D.C.: Office of Under Secretary of Defense.

Department of Defense. (2007, April 23). DoD architecture framework (DODAF), Version 1.5: Volume II, product descriptions. Retrieved March 13, 2009, from http://www.defenselink.mil/cio-nii/docs/DoDAF_Volume_II.pdf

Dodge, M. E. M. (1865). *Hans Brinker, Or the silver skates*. Retrieved on April 14, 2009, from http://books.google.com/books?id=8TsRAAAAYAAJ&pg=PA158&ots=PlbleHItP2&dq=%22hans+brinker+or+The+Silver+Skates:+A+Life+in+Holland%22&output=html

Dooley, K. (1996). Complex adaptive systems: A nominal definition.Retrieved January 5, 2008, from http://www.eas.asu.edu/˜kdooley/casopdef.html

Dreifus, C. (2006, May 14). Earth science meets social science in the study of disasters, *New York Times*. p. F2.

Epstein, S. (2006). Unexampled events, resilience and PRA. In E. Hollnagel & E. Rigaud (Eds.), *Proceedings of the Second Resilience Engineering Symposium*, (pp. 105–116). 8–10 November, Juan-les-Pins, France.

Eyles, D. (2004). *Tales from the lunar module guidance computer*. 27th Annual Guidance and Control Conference of the American Astronautical Society, (paper AAS-04 064). Breckenridge, CO.

Federal Aviation Administration. (2007a). Airworthiness standards: Transport category airplanes. *Federal Aviation Regulation. FAR 25.1309: Equipment, systems and installations*.

Federal Aviation Administration. (2007b). Airplane performance regulation in icing conditions: Final rule. Federal Aviation Regulation, 14 CFR Part 25. *Federal Register (72), 44656-44669*.

Federal Aviation Administration. (1997). *Aviation safety data accessibility study index: Analysis and interpretation of safety data*. Retrieved on March 29, 2009, from http://www.nasdac.faa.gov/aviation_studies/safety_data/analysis.html

Federal Aviation Administration (2004). Safety and security extensions for capability maturity models. Ibrahim, L., et al. (Eds.) Washington, D.C.: Retrieved March 20, 2009, from https://buildsecurityin.us-cert.gov/swa/downloads/SafetyandSecurityExt-Sep2004.pdf

Federal Aviation Administration (FAA) (2008) Code of Federal Regulations (CFR) Final Rule. Reduction of fuel tank flammability in transport category airplanes. July 21. Retrieved from http://www.airweb.faa.gov/Regulatory_and_Guidance_Library/rgFinalRule.nsf/0/5FA0E184CDEF9FE7862574C90055EE05?OpenDocument&Highlight=25-125 August 3, 2009

Federal Aviation Administration (FAA) (2008b). 14 CFR Part 39. Airworthiness Directive; Boeing Model 737 Aircraft. Retrieved on August 4, 2009 from http://rgl.faa.gov/Regulatory_and_Guidance_Library/rgad.nsf/0/9da618efecdae41c862574fd0050510b/$FILE/2008-23-07.pdf

Federal Aviation Administration (FAA) (2009) Airworthiness Directives. Retrieved on August 5, 2009 from http://www.airweb.faa.gov/Regulatory_and_Guidance_Library/rgAD.nsf/Frameset?OpenPage

Feynman, R. (1988). An outsider's inside view of the Challenger inquiry. *Physics Today*, *41*, 2626-37ndash;37.

Flin, R. (2006). Erosion of managerial resilience: From Vasa to NASA. In E. Hollnagel, D. Woods, & N. Leveson (Eds.). *Resilience engineering: Concepts and precepts*. (pp. 223–233). Aldershot, UK: Ashgate.

Flin, R. (2007, June 26). *Managerial decision making: Counterbalancing risks between production and safety* [PowerPoint slides]. Presentation at the Industrial Psychology Research Centre, University of Aberdeen, Aberdeen, Scotland.

Freedman, D. H. (2006). The future of NASA. *Discover*, 34–42.

Garbin, D., & Shortle, J. (2007). Measuring resilience in network-based infrastructures. In *Critical thinking: Moving from infrastructure protection to infrastructure resilience* (pp. 73–86). CIP [Critical Infrastructure Protection] Program Discussion Paper Series, George Mason University.

GCOR (General code of operating rules) (3rd Ed.) (1994). *Train delayed within a block*. Section 9.9. Retrieved on March 8, 2009, from http://www.trainweb.com/gcor/blocks.html#9.9

Global Earth Observation System of Systems (GEOSS). (2006). Working Group of the International Council on Systems Engineering (INCOSE). Retrieved on March 8, 2009, from http://www.incose.org/practice/techactivities/wg/geoss/

Gottlieb, J. (2008a, December 5).Crash puts rail line chief in spotlight. *Los Angeles Times*, p. B4.

Gottlieb, J. (2008b, September 29). Firm that employed Metrolink engineer had other problems. *Los Angeles Times*, p. B1.

Gottlieb, J. (2008c, October 11). Metrolink safety panel appointed. *Los Angeles Times*, p. B6.

Gottlieb, J. (2008d). Metrolink is pairing up engineers. *Los Angeles TImes*, p. B1.

Government Printing Office (GPO) (2009). Electronic Code of Federal Regulations (e-CFR). Title 14—Aeronautics and Space. Part 39—Airworthiness Directives. Retrieved on August 5, 2009 from http://ecfr.gpoaccess.gov:80/cgi/t/text/text-idx?-c=ecfr&tpl=/ecfrbrowse/Title14/14cfr39_main_02.tpl

GRA Incorporated. (2007). *Economic values for FAA investment and regulatory decisions, a guide : Final report.*

Gribble, S. (2001). Robustness in complex systems. *Proceedings of the Eighth Workshop on Hot Topics in Operating Systems*, (pp. 21–26). IEEE Computer Society.

Grote, G. (2006). Rules management as source for loose coupling in high-risk systems. *Proceedings of the Second Symposium on Resilience Engineering*, (pp. 116–124). 8–10 November, Juan-les-Pins, France, November.

Guikema. S. (2009). Infrastructure design issues in disaster-prone regions. *Science, 323*, 13021302-1303ndash;1303.

Hale, A., Guldenmund, F., & Goossens, L. (2006). Auditing resilience in risk control and safety management systems. In E. Hollnagel, D. Woods, & N. Leveson (Eds.). *Resilience engineering: Concepts and precepts.* (pp. 389–314). Aldershot, UK: Ashgate.

Hamill, M. (2006). *Systems engineering: What technical management means.* Presentation to the Los Angeles chapter of INCOSE.

Hardman, N. (2008). What systems engineers need to know about human-computer interaction. *Insight*, 11(2), 19–22.

Haynes, A. (1991). *The crash of United flight 232.* Presentation to NASA-Dryden.

Heisel, W., & Weikel, D. (2007, October 28). Shortages added fuel to O.C. [Orange County] fire. *Los Angeles Times*, pp. A1, A32.

Hennessey-Fiske, M., Connell, R., & Lopez, R. (2008a, October 4). Red light on train wreck disputed. *Los Angeles Times*, p. B1.

Herkströter, C., Moody-Stuart, M., & Steel, G. (1999). Strategic transformation at Royal Dutch/Shell. In Senge, P., et al. *The dance of change: The challenges to sustaining momentum in learning organizations* (pp. 523–28). New York: Doubleday.

Hollnagel, E. (2006). Resilience—the challenge of the unstable. In E. Hollnagel, D. Woods, & N. Leveson (Eds.). *Resilience engineering: Concepts and precepts.* (pp. 9–14). Aldershot, UK: Ashgate.

Hollnagel, E., & Rigaud, E. (Eds.) (2006). *Proceedings of the second Resilience Engineering Symposium.* 8–10 November, Juan-les-Pins, France.

Hollnagel, E., Woods, D., & Leveson, N. (Eds.). (2006). *Resilience engineering: Concepts and precepts.* Aldershot, UK: Ashgate.

Hughes, Stephen (1997). *The Texas City Disaster, 1947*, University of Texas Press.

Hymon, S. (2008a, October 5). Cellphone ban imposed on all train operators. *Los Angeles Times*, p. B1.

Hymon, S. (2008b, September 26). Metrolink to consider safety gear. *Los Angeles Times*, p. B1.

Hymon, S., & Connell, R. (2008, September 25). Train safety improvements are proposed. *Los Angeles Times*, p. B4.

Hymon, S., & Dizikes, C. (2008, September 24). Accord reached on rail bill. *Los Angeles Times*, p.B1.

ISO/IEC 15288. Systems engineering — System life cycle processes. International Organization for Standardization. Geneva, Switzerland: International Organization for Standardization, 2002.

International Council on Systems Engineering (INCOSE). (1998). Fellows' consensus on systems engineering. *International Council on Systems Engineering (INCOSE).* Retrieved on March 27, 2009, from http://www.incose.org/practice/fellows consensus.aspx

International Council on Systems Engineering (INCOSE). (2006). *Systems engineering handbook: A guide for system life cycle processes and activities* (Version 3). Haskins, C. (Ed.) Seattle, WA.

Jackson, D. (2006). Dependable software by design. *Scientific American, 294*, 68–75.

Jackson, S. (1997). *Systems engineering for commercial aircraft.* Aldershot, UK: Ashgate.

Jackson, S. (2002). Organizational safety: A systems engineering perspective. *Proceedings of the INCOSE International Symposium* (paper 4.2.2). 28 July–1 August, Las Vegas, NV.

Jackson, S. (2007). A multidisciplinary framework for resilience to disasters and disruptions. *Journal of Design and Process Science, 11*, 91–108.

Jackson, S. (2008). *The California firestorms of 2007.* Panel presentation to the International Council on Systems Engineering (INCOSE) Symposium. 18-June, Utrecht, the Netherlands.

Jackson, S., & Hann, S. (2004). Attributes of a managerial and organizational infrastructure to enable safe systems. *Proceedings of the INCOSE International Symposium* (paper 6.5.1). 2020-24ndash;24 June, Toulouse, France.

Jackson, S., Erlick, K., & Gutierrez, J. (2006). The science of organizational psychology applied to mission assurance. *Conference on Systems Engineering Research* (paper 102). 77-8ndash;8 April, Los Angeles, CA.

Kaslow, D. (2004). Factors contributing to space systems failures and success. *Proceedings of the 13th Annual International Council on Systems Engineering (INCOSE)*, (pp. 992–1006). 2020-24ndash;24 June, Toulouse, France.

Ketchum, M. (n.d.). Mark Ketchum's bridge collapse page. Retrieved on March 8, 2009, from http://www.ketchum.org/bridgecollapse.html

Kopp, C. (1993, September 7). Anatomy of a mistake: The tragic death of Jessica Santillan. *60 Minutes special report.* New York: Columbia Broadcasting System (CBS).

Laddaga, R., & Robertson, P. (n.d.). Model based diagnosis in self adaptive software. Retrieved on March 29, 2009, from http://monet.aber.ac.uk:8080/monet/docs/pdf_files/dx03/ladd_dx03.pdf

Larson, E. (2000). *Isaac's storm: The time, and the deadliest hurricane in history.* New York: Vantage Books.

Leary, W. (2008). Wielding a cost-cutting ax, and often, at NASA. *New York Times.* p. F4.

Leveson, N. (1995). *Safeware: System safety and computers.* Reading, MA: Addison Wesley.

Leveson, N. (2002a). A new approach to system safety engineering. MIT. Manuscript in preparation. Retrieved March 18, 2009, from http://ocw.mit.edu/NR/rdonlyres/

Aeronautics-and-Astronautics/16-358JSpring-2005/7A17C38C-F622-4244-ABF0-5BD2B768661C/0/book2.pdf

Leveson, N. (2002b). *System safety engineering: Back to the future*. Manuscript in preparation. Retrieved March 8, 2009, from http://sunnyday.mit.edu/book2.pdf

Leveson, N. (2004a). A new accident model for engineering safer systems. *Safety Science*, *42*, 237–270.

Leveson, N. (2004b). Model-based analysis of socio-technical risk. *MIT Engineering Systems Division Working Paper Series*. Retrieved March 13, 2009, from http://esd.mit.edu/wps/esd-wp-2004-08.pdf

Leveson, N. (2008). System safety engineering for software-intensive systems [Power Point slides]. Presentation at Boeing, Long Beach, CA.

Leveson, N., et al. (2005). *Modeling, analyzing and engineering NASA's safety culture: Phase 1 Final Report*, September 2004 to February 2005. Retrieved March 13, 2009, from http://sunnyday.mit.edu/Phase1-Final-Report.pdf

Leveson, N., et al. (2006). Engineering resilience into safety-critical systems. In E. Hollnagel, D. Woods, & N. Leveson (Eds.). *Resilience engineering: Concepts and precepts*. (pp. 95–122). Aldershot, UK: Ashgate.

Lintern, G. (2003). Tyranny in rules, autonomy in maps: Closing the safety management loop. *Proceedings of the Twelfth International Symposium on Aviation Psychology*, (pp. 719–24). 1414-17ndash;17 April, Dayton, OH.

Liu, J., et al. (2007). Complexity of coupled human and natural systems. *Science*, 317, 1513–1516.

Lopez, R., & Connell, R. (2008a, December 26). Experts see danger in signal gaps. *Los Angeles Times*, p. B1.

Lopez, R., & Connell, R. (2008b, December 5). Signal was green, says crewman. *Los Angeles Times*, p. B1.

Lopez, R., & Hymon, S. (2008a, October 7). Train engineer sent text message just before crash. *Los Angeles Times*, p. B1.

Lopez, R., & Hymon, S. (2008b, September 18). Train's engineer received, sent text messages on duty. *Los Angeles Times*, p. B1.

Lopez, R., & Hymon, S. (2009, January 7). Firm knew Metrolink engineer sent text messages, lawyers, say. *Los Angeles Times*, p. B3.

Lopez, R., & Oldham, J. (2008, September 16). Warning signals were working, officials say. *Los Angeles Times*, p. B1.

Lopez, R., & Weikel, D. (2008, September 17). Metrolink balked at safety upgrade's cost. *Los Angeles Times*, p. A1.

Lopez, R., Connell, R., & Hymon, S. (2008, November 22). Metrolink train ran red light. *Los Angeles Times*, p. B1.

Madni, A. (2007). *Design for resilience* [PowerPoint slides]. Presentation at Intelligent Systems Technologies, Inc. (ISTI), Los Angeles, CA.

Madni, A, Ahlers, R., & Chu, Y. (1987). Knowledge-based simulation: An approach to intelligent opponent modeling for training tactical decision making. *Proceedings of the 9th Interservice/Industry Training Systems Conference*, (pp. 179–183). 30 November-2 December, Washington, D.C.

Madni, A., & Jackson, S. (2009). Towards a conceptual framework for resilience engineering. *IEEE Systems Journal*. In Press.

Malakis, S., & Kontogiannis, T. (2008). Cognitive strategies in emergency and abnormal situations training: Implications for resilience in air traffic control. *Proceedings of the Third Symposium on Resilience Engineering*, 28–30 October, Antibes, Juan-les-Pins, France.

Mallak, L. (1998). Measuring resilience in health care provider organizations, *Health Manpower Management*, *24*, 148–154.

Marais, K., Dulac, N., & Leveson, N. (2004). Beyond normal accidents and high reliability organizations: The need for an alternative approach to safety in complex systems. *Engineering Systems 2004 Symposium. MIT*. Retrieved March 13, 2009, from http://esd.mit.edu/symposium/pdfs/papers/marais-b.pdf

Maier, M., & Rechtin, E. (2009). *The art of systems architecting* (3rd ed.). Boca Raton, FL: CRC Press.

Mathews, M. (Ed.) (1951). *A dictionary of Americanisms on historical principles* (Vol. *1*). Chicago, IL: University of Chicago Press.

Matthews, R., & Romanowski, M. (2004). *Cooperative efforts are driving down the accident rate* [PowerPoint slides]. US/Europe International Safety Conference, Philadelphia, June 7–11.

Maxwell, J. (2009). *Observations of the resilience architecture of the firefighting and emergency response infrastructure* [Research Paper]. University of Southern California, Los Angels, CA. February 9.

McCarthy, J. (2007). From protection to resilience: Injecting 'moxie' into the infrastructure security continuum. In *Critical thinking: Moving from infrastructure protection to infrastructure resilience*, (pp. 1–7). CIP [Critical Infrastructure Protection] Program Discussion Paper Series, George Mason University.

McFadden, R. (2009, January 16). All 155 aboard safe as crippled jet crash-lands in Hudson. *New York Times*, p. A1.

McKinley, J., & Wald, M. (2008, September 19). California bans texting by operators of trains. *New York Times*, p. A10.

Mendoa, D., & Wallace, W. (2006). Adaptive capacity: Electric power restoration in New York City following the 11 September 2001 attacks. In E. Hollnagel & E. Rigaud(Eds.), *Proceedings of the Second Resilience Engineering Symposium*, (pp. 209-19). 8-10 November, Juan-les-Pines, France.

Meshkati, N., & Osborn, J. (2008, September 17). Rail safety and the excuse of human error. *Los Angeles Times*, p. A23.

MIL-STD-882. Standard practice for system safety. Department of Defense. 1993.

Moody, J.,et al. (1997). *Metrics and case studies for evaluating engineering designs*. Upper Saddle River, NJ: Prentice Hall.

Morrison, P. (2008, September 18). She told the truth and lost her job. *Los Angeles Times*, p. A25.

Murphy, K. (2006, October 31). What pilots can teach hospitals about patient safety. *New York Times*, pp. D6, D10.

Nadler, G., & Chandon, W. (2004). *Smart questions : Learn to ask the right questions for powerful results*. San Francisco, CA: Jossey-Bass.

NASA Jet Propulsion Laboratory. California Institute of Technology. (1997). *NASA Mars climate orbiter web site*. Retrieved on March 8, 2009, from http://mars.jpl.nasa.gov/msp98/orbiter/

National Aeronautics and Space Administration (NASA). (1997). Safety, reliability, m aintainability and quality provisions for the Space Shuttle program. Report NSTS, 5300.4 (1D-2).

National Aeronautics and Space Administration (NASA). (2008). *Process based mission assurance (PBMA) knowledge management system*. Retrieved March 13, 2009, from http://pbma.nasa.gov/pbma_main

National Transportation Safety Board. (2000). *Aircraft accident report, in-flight breakup over the Atlantic Ocean, Trans World Airline Flight 800, July 17, 1996*. NTSB Number AAR-00/03; NTIS Number PB2000-910403. Washington, D.C.: United States National Safety Transportation Board.

National Transportation Safety Board (NTSB) (1990). Safety recommendation. A-9090-167ndash;167 through 175. December 14. Retrieved from http://www.ntsb.gov/recs/letters/1990/A90_167_175.pdf on August 4, 2009

Nicolescu, B. (2007). Transdisciplinarity as methodological framework for going beyond the science-religion debate. *The Global Spiral*, *8*. Retrieved March 27, 2009, from http://www.metanexus.net/magazine/tabid/68/id/10013/Default.aspx

Nolte, J. (2008). *Immediate response and long term recovery from the 9-11 terrorist attack: What should we do differently in a future attack?* [PowerPoint slides]. Panel presentation to the International Council on Systems Engineering (INCOSE) Symposium, 18 June, Utrecht, The Netherlands.

Norman, D. (1982). Steps toward a cognitive engineering. *Proceedings of the SIGCHI (Special Interest Group on Computer Human Interaction) Conference on Human Factors in Computing Systems*, (pp. 378–382). 15-17 March, Gaithersburg, MD.

Okada, Y. (2003). Analysis of potential human error in hospital work. In D. Harris, et al. *Human-centered computing: Cognitive, social and ergonomic aspects*, Vol. 3. (pp. 532–536). Mahwah, NJ: Lawrence Erlbaum Associates.

Paté-Cornell, M. E. (1990). Organizational aspects of engineering system safety: The case of offshore platforms. *Science, 250*, 1210–1217.

Paté-Cornell, M. E., & Fischbeck, P. (1994). Risk management for the tiles of the space shuttle interface. *Interfaces, 24*, 64–86.

Patterson, E., Woods, D., Cook, R., & Render, M. (2007). Collaborative cross-checking to enhance resilience.*Cognition, Technology & Work*, *9*, 155–162.

Perelman, L. (2007). Shifting security paradigm—toward resilience. In *Critical thinking: Moving from infrastructure protection to infrastructure resilience*, (pp. 23–48). CIP [Critical Infrastructure Protection] Program Discussion Paper Series, George Mason University.

Perrow, C. (1999). *Normal accidents: Living with high-risk technologies*. New York: Basic Books.

Pommerening, C. (2007). Resilience in organizations and systems: Background and trajectories of an emerging paradigm. In *Critical thinking: Moving from infrastructure protection to infrastructure resilience*, (pp. 9–21). CIP [Critical Infrastructure Protection] Program Discussion Paper Series, George Mason University.

Reason, J. (1990). *Human error*. Cambridge, UK: Cambridge University Press.

Reason, J. (1997). *Managing the risks of organizational accidents.* Aldershot, UK: Ashgate.

Rechtin, E. (1991). *Systems architecting: Creating and building complex systems.* Englewood Cliffs, NJ: Prentice Hall.

Rechtin, E. (2000). *Systems architecting of organizations: Why eagles can't swim.* Boca Raton, FL: CRC Press.

Resilience Alliance. (2008). *Resilience.* Retrieved December 30, 2007, from http://www.resalliance.org/576.php

Richards, R., Ross, M., Hastings, D., & Rhodes, D. (2007). Design principles for survivable system architecture. *Proceedings of the 1st IEEE Systems Conference,* 9, 1–9. 9–13. Honolulu, HI.

Rohrlich, T. (2008). Train crash's roots run deep. *Los Angeles Times*, pp. A1, A22–23.

Rooney, J., Vanden Heuvel, L., & Lorenzo, D., Stoecklein, M., & Christensen Stoecklein, M., & Christensen, M.M. (2002). Reduce human error. *Quality Progress, 35*, 27–36.

Sage, A. (2008). *Federated system of systems management* [PowerPoint slides]. Presentation to the Conference on Systems Engineering Research, 4–5 April, Los Angeles, CA.

San Francisco Fire Department (SFFD). (1983). *Manual for standard practices.*

Scalingi, L. (2007). Moving beyond critical infrastructure protection to disaster resilience. In *Critical thinking: Moving from infrastructure protection to infrastructure resilience*, (pp. 49–71). CIP [Critical Infrastructure Protection] Program Discussion Paper Series, George Mason University.

Schwartz, J. (2008, January 9). With U.S. help, private space companies press their case: Why not us? *New York Times*, p. D4.

Senge, P. (1999). The growth processes of profound change. In Senge, P., et al. *The dance of change: The challenges to sustaining momentum in learning organizations* (pp. 42–57). New York: Doubleday.

Senge, P., et al. (1999). *The dance of change: The challenges to sustaining momentum in learning organizations.* New York: Doubleday.

Sheard, S. (2006). *What you will need to know about chaos, complexity, and complex adaptive systems to do systems engineering well into the 21st century.* International Council on Systems Engineering (INCOSE) tutorial, 9-13 July, Orlando, FL.

Stephan, K. (2007). We've got to talk: Emergency communications and engineering ethics. *IEEE Technology & Society Magazine, 26*, 42-48.

Stephens, H. (1997). *The Texas City disaster, 1947.* Austin, TX: University of Texas Press.

Technologies Strategies International (TSI) (2009). *The systems approach.* Retrieved March 8, 2009, from http://tsicanada.com/systems_approach.htm

The Nestlé management and leadership principles. (2003). Retrieved on March 8, 2009, from http://www.nestle.com/Resource.axd?Id=1353AE38-2F44-4F5F-A31F-72D57EE0CF35

U.S. Department of Transportation (DOT).(1994).*Technical advisory: The cost of motor vehicle accidents, T7570.2.*

United States Air Force. (2000). *Operational safety, suitability and effectiveness plan.* Air Force Instruction 63-1201.

University of Washington. (2009). Systems theory. *College of Forest Resources.* Retrieved on March 8, 2009 from http://silvae.cfr.washington.edu/ecosystem-management/Systems.html

Utts, J., & Heckard, R. (2005). *Statistical ideas and methods.* Pacific Grove, CA: California Thomas Brooks Cole.

Van Allen radiation belt. (2008). In Encyclopedia Britannica. Retrieved on March 8, 2009, from http://www.britannica.com/eb/article-9074758/Van-Allen-radiation-belt

Vaughn, D. (1996). *The Challenger launch decision: Risky technology, culture, and deviance at NASA.* Chicago, IL: University of Chicago Press.

Wald, M. (2003, July 21). Study suggests space agency should consider mimicking Navy's safety techniques. *New York Times*, p. A13.

Wald, M. (2005, September 1). More planes too close than reports had listed. *New York Times*, p. A21.

Wald, M., & Schwartz, J. (2003, August 4). Shuttle inquiry uncovers flaws in communications. *New York Times*, p. A9.

Watt, K. (1974). *The Titanic effect: Planning for the unthinkable.* Stamford, CT: Sinauer Associates.

Wegner, E. (1998). *Communities of practice: Learning, meaning and identity.* Cambridge, UK: University of Cambridge.

Weick, K., & Sutcliffe, K. (2001). *Managing the unexpected: Assuring high performance in an age of complexity.* San Francisco, CA: Jossey-Bass.

Weikel, D., Connell, R., & Lopez, R. (2008, September 18). Engineer's split shift is probed. *Los Angeles Times*, p. A1.

Werner, P. (2001). Assessing an organization's culture. *Proceedings of the Eleventh Annual Symposium of the International Council on Systems Engineering*, (paper 1271). 1–5 July, Melbourne, Australia.

Westrum, R. (2006a). A typology of resilience situations. In E. Hollnagel, D. Woods, & N. Leveson (Eds.). *Resilience engineering: Concepts and precepts.* (pp. 55–65). Aldershot, UK: Ashgate.

Westrum, R. (2006b). All coherence gone: New Orleans as a resilience failure. In E. Hollnagel & E. Rigaud (Eds.), *Proceedings of the Second Resilience Engineering Symposium*, (pp. 333–41). 8–10 November, Juan-les-Pins, France.

White, J. (2006). The strongest handshake in the world. *Invention and Technology*, 21, 51–54.

White House (1993). Executive order 12866. Regulatory planning and review. September 30. Retrieved on August 5, 2009 from http://www.whitehouse.gov/omb/inforeg/eo12866/eo12866_amended_01-2007.pdf

Wilbur, C. (n.d.). Holistic method. *University of Notre Dame.* Retrieved on March 8, 2009, from http://www.nd.edu/˜cwilber/pub/recent/edpehol.html

Woods, D. (2005). *Creating foresight: Lessons for enhancing resilience in Columbia* [draft article]. Retrieved on March 18, 2009, from http://csel.eng.ohio-state.edu/woods/space/Create%20foresight%20Col-draft.pdf

Woods, D. (2006). How to design a safety organization: Test case for resilience engineering. In E. Hollnagel, D. Woods, & N. Leveson (Eds.). *Resilience engineering: Concepts and precepts*. (pp. 315–26). Aldershot, UK: Ashgate.

Woods, D. (2006a). Epilogue: Resilience engineering precepts.In E. Hollnagel, D. Woods, & N. Leveson (Eds.). *Resilience engineering: Concepts and precepts*. (pp. 347–348). Aldershot, UK: Ashgate.

Woods, D. (2006b). Essential characteristics of resilience.In E. Hollnagel, D. Woods, & N. Leveson (Eds.). *Resilience engineering: Concepts and precepts*. (pp. 21–34). Aldershot, UK: Ashgate.

Woods, D. (2006c). Laws that govern joint cognitive systems at work. *Ohio State University*. Retrieved March 26, 2009, from http://csel.eng.ohio-state.edu/productions/laws/laws_flash1.swf

Woods, D. (2007). *Navigating complexity* [video file]. Lecture at Almaden Research Center, CA, April 11-12. Retrieved on March 8, 2009, from http://csel.eng.ohio-state.edu/productions/ibm/

Woods, D., & Cook, R. (2006). Incidents—markers of resilience or brittleness?In E. Hollnagel, D. Woods, & N. Leveson (Eds.). *Resilience engineering: Concepts and precepts*. (pp. 69–76). Aldershot, UK: Ashgate.

Woods, D.,Wreathall, J., & Anders, S. (2006). Stress-strain plots as a model of an organization's resilience. *Proceedings of the Second Symposium on Resilience Engineering*, (pp. 342–349). 8-10 November, Juan-les-Pins, France.

Wright, L., & Van der Schaaf, T. (2004). Accident versus near-miss causation: A critical review of the literature, an empirical test in the UK railway domain, and their implications for other sectors. *Journal of Hazardous Materials*, 111, 105–110.

Zachman, J. (2007). *Enterprise design objectives: Complexity and change* [PowerPoint slides]. Retrieved March 26, 2009, from http://www.ndia.org/DoDEntArchitecture/Documents/HalfDay092Up.pdf

Zarboutis, N., & Wright, P. (2006). Using complexity theories to reveal emerged patterns that erode the resilience of complex systems. *Proceedings of the Second Symposium on Resilience Engineering*, (pp. 359–368). 8–10 November, Juan-les-Pins, France.

Index

Architecting Resilient Systems: Accident Avoidance and Survival and Recovery from Disruptions.
By Scott Jackson

WILEY SERIES IN SYSTEMS ENGINEERING AND MANAGEMENT

Andrew P. Sage, Editor

ANDREW P. SAGE and JAMES D. PALMER
Software Systems Engineering

WILLIAM B. ROUSE
Design for Success: A Human-Centered Approach to Designing Successful Products and Systems

LEONARD ADELMAN
Evaluating Decision Support and Expert System Technology

ANDREW P. SAGE
Decision Support Systems Engineering

YEFIM FASSER and DONALD BRETTNER
Process Improvement in the Electronics Industry, Second Edition

WILLIAM B. ROUSE
Strategies for Innovation

ANDREW P. SAGE
Systems Engineering

HORST TEMPELMEIER and HEINRICH KUHN
Flexible Manufacturing Systems: Decision Support for Design and Operation

WILLIAM B. ROUSE
Catalysts for Change: Concepts and Principles for Enabling Innovation

LIPING FANG, KEITH W. HIPEL, and D. MARC KILGOUR
Interactive Decision Making: The Graph Model for Conflict Resolution

DAVID A. SCHUM
Evidential Foundations of Probabilistic Reasoning

JENS RASMUSSEN, ANNELISE MARK PEJTERSEN, and LEONARD P. GOODSTEIN
Cognitive Systems Engineering

ANDREW P. SAGE
Systems Management for Information Technology and Software Engineering

ALPHONSE CHAPANIS
Human Factors in Systems Engineering

YACOV Y. HAIMES
Risk Modeling, Assessment, and Management, Third Edition

DENNIS M. BUEDE
The Engineering Design of Systems: Models and Methods, Second Edition

ANDREW P. SAGE and JAMES E. ARMSTRONG, Jr.
Introduction to Systems Engineering

WILLIAM B. ROUSE
Essential Challenges of Strategic Management

YEFIM FASSER and DONALD BRETTNER
Management for Quality in High-Technology Enterprises

THOMAS B. SHERIDAN
Humans and Automation: System Design and Research Issues

ALEXANDER KOSSIAKOFF and WILLIAM N. SWEET
Systems Engineering Principles and Practice

HAROLD R. BOOHER
Handbook of Human Systems Integration

JEFFREY T. POLLOCK AND RALPH HODGSON
Adaptive Information: Improving Business Through Semantic Interoperability, Grid Computing, and Enterprise Integration

ALAN L. PORTER AND SCOTT W. CUNNINGHAM
Tech Mining: Exploiting New Technologies for Competitive Advantage

REX BROWN
Rational Choice and Judgment: Decision Analysis for the Decider

WILLIAM B. ROUSE AND KENNETH R. BOFF (editors)
Organizational Simulation

HOWARD EISNER
Managing Complex Systems: Thinking Outside the Box

STEVE BELL
Lean Enterprise Systems: Using IT for Continuous Improvement

J. JFRRY KAUFMAN AND ROY WOODHEAD
Stimulating Innovation in Products and Services: With Function Analysis and Mapping

WILLIAM B. ROUSE
Enterprise Tranformation: Understanding and Enabling Fundamental Change

JOHN E. GIBSON, WILLIAM T. SCHERER, AND WILLAM F. GIBSON
How to Do Systems Analysis

WILLIAM F. CHRISTOPHER
Holistic Management: Managing What Matters for Company Success

WILLIAM B. ROUSE
People and Organizations: Explorations of Human-Centered Design

GREGORY S. PARNELL, PATRICK J. DRISCOLL, AND DALE L. HENDERSON
Decision Making in Systems Engineering and Management

MO JAMSHIDI
System of Systems Engineering: Innovations for the Twenty-First Century

ANDREW P. SAGE AND WILLIAM B. ROUSE
Handbook of Systems Engineering and Management, Second Edition

JOHN R. CLYMER
Simulation-Based Engineering of Complex Systems, Second Edition

KRAG BROTBY
Information Security Governance: A Practical Development and Implementation Approach

JULIAN TALBOT AND MILES JAKEMAN
Security Risk Management Body of Knowledge

SCOTT JACKSON
Architecting Resilient Systems: Accident Avoidance and Survival and Recovery from Disruptions